Innovation and Technology in Europe

THE NATURE OF INDUSTRIALIZATION

Series editors: *Peter Mathias and John A. Davis*

This series is based on the graduate seminars in economic history that have been sponsored by the *Istituto Italiano per gli Studi Filosofici* (Naples) and held annually at the *Centre for Social History* in the University of Warwick.

Published

Volume 1 *The First Industrial Revolutions*
Volume 2 *Innovation and Technology in Europe*

In preparation

Volume 3 *Labour and Enterprise from the Eighteenth Century to the Present*
Volume 4 *International Trade and British Economic Growth 1750–1990*
Volume 5 *Agriculture and Industrialization 1700–1980*
Volume 6 *Science and Industrial Technology 1850–1990*

Innovation and Technology in Europe

From the Eighteenth Century to the Present Day

Edited by Peter Mathias and John A. Davis

BLACKWELL
Oxford UK & Cambridge USA

Copyright © Basil Blackwell 1991

First published 1991

Basil Blackwell Ltd
108 Cowley Road, Oxford, OX4 1JF, UK

Basil Blackwell, Inc.
3 Cambridge Center
Cambridge, Massachusetts 02142, USA

All rights reserved. Except for the quotation of short passages for the purposes of criticism and review, no part of this publication may be reproduced, stored in a retrieval system, or transmitted, in any form or by any means, electronic, mechanical, photocopying, recording or otherwise, without the prior permission of the publisher.

Except in the United States of America, this book is sold subject to the condition that it shall not, by way of trade or otherwise, be lent, re-sold, hired out, or otherwise circulated without the publisher's prior consent in any form of binding or cover other than that in which it is published and without a similar condition including this condition being imposed on the subsequent purchaser.

British Library Cataloguing in Publication Data
A CIP catalogue record for this book is available from the British Library

Library of Congress Cataloging in Publication Data
Innovation and technology in Europe: from the eighteenth century to the present day/edited by Peter Mathias and John A. Davis.
p. cm.—(The Nature of industrialization)
Includes bibliographical references and index.
ISBN 0-631-16797-8
1. Technological innovations—Economic aspects—Europe—History.
I. Mathias, Peter.   II. Davis, John Anthony.   III. Series.
HC240.9.T4154   1991
338'.064'094—dc20   90-19962
                                                                              CIP

Typeset in 10 on 12pt Garamond by Wearside Tradespools, Fulwell, Sunderland
Printed in Great Britain by Hartnolls Ltd., Cornwall

Contents

Preface		vii
Editors' Introduction		1
1	The Mainsprings of Technological Progress in Europe 1750–1850 *Patrick O'Brien*	6
2	Resources and Technology *Peter Mathias*	18
3	Revisions and Revolutions: Technology and Productivity Change in Manufacture in Eighteenth-century England *Maxine Berg*	43
4	The Constraints of a Proto-industrial Society on the Development of Heavy Industry: the Case of Coal-mining in the South-east of France 1773–1791 *Gwynne Lewis*	65
5	Innovation in an Industrial Late-comer: Italy in the Nineteenth Century *John A. Davis*	83
6	Technical and Structural Factors in British Industrial Decline 1870 to the Present *Derek H. Aldcroft*	107
7	Crisis and Continuity: Innovation in the British Automobile Industry 1896–1986 *Richard Whipp*	120
8	Technology and the Export of Industrial Culture: Problems of the German–American Relationship 1900–1960 *Volker R. Berghahn*	142

9 Technological Diffusion: the Viewpoint of Economic
  Theory
  Paul L. Stoneman 162
Notes on Contributors 185
Index 186

Preface

This volume is based on papers that were first delivered at the second economic history summer school held at the Centre for Social History in the University of Warwick in July 1986, in collaboration with the Italian Institute for Philosophical Studies. The editors wish to take this opportunity to thank the President of the Institute, Avvocato Gerardo Marotta, and Professor Luigi De Rosa for their continuing and generous support.

Editors' Introduction

How and why does technological change occur? What are the conditions that favour first the development and then the application of new technologies? How does inventiveness become translated into innovation? Why has successful innovation occurred in certain places and at certain times, but not in others? Is innovation a conscious, deliberate, planned and cumulative process – or is it accidental and discontinuous? Has technology been the prime mover of economic growth and industrialization – or is the technological process itself influenced, or even determined, by other forces and circumstances? What is the relationship between technological change and the social, political and cultural contexts in which it takes place?

These are the questions that are addressed in the essays that make up this volume. Although they have long held a central place in the study of modern industrialization and economic growth, many of the most fundamental aspects of the process of technological change are still shrouded in uncertainties. This volume cannot seek to dispel those uncertainties, but it does seek to offer a broad introduction to some of the issues and problems they pose. The approach is not comprehensive, but draws selectively on the experience of technological change in Western Europe since the eighteenth century to illustrate the complexity of the factors which have influenced how new technologies are created and applied.

Even within the geographical area chosen, space and selection mean that certain problems receive more attention than others, while there are also some omissions. Most of the essays are concerned with technological change in the development of manufacturing industries, and for a fuller understanding of the role of technological innovation in the process of economic growth greater account needs to be taken of other fields of application, especially agriculture, communications and transport, services and social utilities. For reasons of space the impact of the scientific revolutions of the late nineteenth century on applied science and industrial technology is only touched on here, and this important topic will be the subject of a separate volume to appear later in the series.

Within these limits, however, the volume explores a series of closely related themes in different contexts of time and place. Its starting point is current debate on the origins and nature of the process of technological change and the ways in which recent research has begun to set new questions and undermine older certainties. Many previously accepted interpretations and commonplaces no longer hold good. Gone are the days when the economic historian could comfortably attribute England's precocious industrial primacy in the late eighteenth century, for example, to the heroic clutch of inventions that launched mechanized spinning, steam power, deeper mining, improved iron-making and new forms of transportation. None of the 'single-cause' explanations of British inventiveness in the eighteenth century has stood the test of sustained investigation, nor do there seem to be valid generalizations linking individual genius, collective knowledge or levels of scientific progress to inventiveness. It is no easier – to remain with the same example – to single out a mono-causal explanation for Britain's relative failure in the twentieth century to maintain competitive levels of technological innovation in so many fields.

The passing of older certainties has revealed not only unanswered questions but also complex methodological problems. One of the most fundamental – and this theme recurs throughout the volume – is that of definition. Technology is neither self-explanatory nor singular. We may say in general that technology provides the means to achieve new forms of production that are increasingly capital intensive and permit ever-higher levels of productivity. But historically that definition is too restricting and technologies have come in a great variety of shapes, sizes and functions. The variables are almost infinite and may assume any number of combinations: a technology may speed up production, improve product quality, reduce product costs, or make possible product innovation; it may be labour-saving or capital-saving, or both, or neither; some technologies increase demand for skilled labour, others have the reverse effect; some technologies have increased demands for capital, others have not.

Recent research has also given greater prominence to the 'knock-on' effects and inter-relatedness of innovations, which often produce quite unexpected secondary effects. As a result, innovation is increasingly seen not as the result of single breakthroughs, but as a more systemic process of continuous technological change, innovation and adaptation at a variety of different levels, each with its own distinctive impact.

Technology in other words is not singular, and it is more useful to think in terms of technologies, just as the impact of technological change varies from sector to sector, from context to context. For this reason technological change cannot be studied except in the specific context within which new technologies are created and applied. Technology cannot even be defined except in relation to historically specific configurations of resources, techniques, methods of production and factor endowments. Only in those contexts is it possible to distinguish between what is new and what is old, to explain how the new is generated, how and why older technologies were – or were not – displaced, and what the consequences were. Only in the context of specific factor endowments and opportunities is it possible to identify and explain the comparative advantage offered by a particular innovation and hence to measure its impact on existing and earlier methods of production.

Questions of context and definition link closely with the problem of measurement. It is obvious that without clear definition there can be no satisfactory evaluation of the pace or scale of technological change. But arbitrary categorization of high and low technology sectors within a given economy may mask real processes of technological change in ways that distort the overall picture. The use of large scale aggregate data in macroeconomic quantitative studies may hide qualitative changes at a sectoral or regional level that might have profound significance for wider economic growth. Economic historians have also begun to question the association between technological innovation and the rise of the newest and largest-scale nineteenth-century industries, arguing from cases like the French silk industry that effective technological development was possible within the framework of smaller scale industries where more traditional forms of organization survived.

It has always been recognized that the dynamic relations between 'high' technology (capital-intensive) and 'low' technology (labour-intensive) were complex – and in different phases of growth, as between different regions, as much complementary as confrontational. Mechanized spinning co-existed with hand-loom weaving; large-scale primary production of metals was associated with small artisan-dominated workshop production of finished goods in the secondary metal industries.

Conventional distinctions between 'high' and 'low' technologies and between new and traditional sectors may be deceptive and Maxine Berg argues that it is mistaken to argue that there was no significant level of technological change in those nineteenth-century British industries that depended largely on the labour of women and children. Although they are conventionally designated as 'low technology sectors', their development was only made possible by technological innovations that radically transformed methods of production and levels of productivity.

The context of technological change is not determined only by the organization of production, the availability of resources, materials, labour and so forth. Political organization, and social and institutional structures have also deeply influenced the process of technological change, and are examined in the essays in this volume on the obstacles to the development of the coal-mining industry in south-eastern France in the late eighteenth century, on technological development in nineteenth-century Italy, and on the role of technology in Britain's industrial decline in the twentieth century.

A more specific, and immediate, context of technological change is the company and the individual firm. In the case of the British motor industry in the twentieth century, examined here by Richard Whipp, the close inter-dependence between technological innovation, company management, strategy and planning procedures becomes very evident. Not only are many different levels of technologies involved within a single product, but they are used for different purposes (engineering and design, for example). If this illustrates the difficulty of isolating a single 'technological factor', it gives added importance to understanding how technological considerations are accommodated in the co-ordination of a company's forward planning and strategy, investment preferences and marketing capacity, as well as in design, research and development operations.

A different dimension of the context of technological innovation is illustrated in Volker Berghahn's essay on German attitudes to American production methods and company organization in the twentieth century. It has long been recognized that technology is never neutral or value free, and this has always been a central theme of the classical studies on the social and political impact of the industrial revolutions. In the case of German industrialists and managers in the twentieth century, Berghahn argues that resistance to American technology was determined by hostility to the political and social values that accompanied mass production. It was only after the Second World War, he concludes, when a new generation of German businessmen began to show greater sympathy for democratic ideals and the consumer society, that this resistance was gradually overcome.

Editors' Introduction

The emphasis on the importance of the specific context of innovation makes generalization difficult, and this – as Patrick O'Brien argues in his opening chapter – widens the gap betwen macroeconomic theories of technological change and the micro-historical circumstances surrounding particular innovations. The historian's focus on the particular and the specific is one reason for this; but the fault also lies – as O'Brien rightly indicates – with the inadequacies of economic theories of innovation that are couched in terms of simple models of supply or demand.

Since these criticisms are underscored at various points throughout the volume, it is fitting and fair to close with the viewpoint of an economist. Paul Stoneman's concluding chapter is directed primarily to economists, but what it has to say should be of real importance to the economic historian as well. Given the specifically delimited issue to which it is addressed – a theoretical approach to the process of technology diffusion – the range and complexity of the variables contained in Stoneman's theoretical model must give economic historians pause to consider whether, in their own analysis of how technology transfers have occurred in the real world of the past, they are even now prepared to take sufficient account of their complexity. On the other hand, the parameters with which Paul Stoneman's analysis are developed quite intentionally omit most of those elements of time, place and circumstance which in a historical context appear to have exercised decisive influences on the processes of technological innovation and economic growth. The gap between the approach of the economist and that of the economic historian is starkly laid bare, and if we are to achieve a better understanding of those processes of technological change that influence so deeply every aspect of the contemporary world it needs to be narrowed.

If this volume demonstrates the multiple and heterogeneous character of the historical contexts of innovation, it does not offer a clear-cut answer to the questions with which we started. To echo Derek Aldcroft's concluding comment, it is not pleasant to end on an inconclusive note. The comfort of past certainties may have gone, but this is mainly because single-cause explanations have been found wanting. Like economists, economic historians are increasingly aware of the complexity of the forces, circumstances and interactions that have historically shaped the processes of innovation. The picture that emerges is often untidy, but it does reveal the highly differentiated nature of the historical experience of innovation.

Peter Mathias
John A. Davis

1
The Mainsprings of Technological Progress in Western Europe 1750–1850

Patrick O'Brien

Introduction

Long run economic growth originates from a multiplicity of sources. But in describing and explaining the economic development of Western Europe from the middle ages through to modern times historians have periodized those ten centuries into three 'epochs', each distinguished and differentiated by a dominant or prime mover behind the more or less rapid rates of expansion experienced by European economies over time. As historians see it, economic progress, where and when it occurred in the medieval epoch, emanated basically from an extension of the agricultural frontier to the east, which formed the basis for increased trade and specialization within a predominantly agrarian society.

From the fifteenth century to the eighteenth century, the extension of markets contingent on the voyages of discovery and on improvements in waterborne transportation has emerged in the historical literature as the focus of European economic development. The epochal innovation that marks out the period from around 1700 onwards can be singled out (to use the words of Simon Kuznets) 'as the extended application of science based technology to the problems of production'.[1] Accepting this quotation as a reasonable one-line depic-

---

[1] S. Kuznets, *Modern Economic Growth* (New Haven, Conn., 1966), p. 9.

tion of the *primum mobile* of economic development over the past three centuries, this chapter will proceed first to define technical progress and then to evaluate several explanations for the origins and also for the changing locations of new technologies which emerged in various European and North American societies over the period 1750 to 1850.

Towards some functional definitions of technological progress

Philosophers and social scientists may appear to be excessively preoccupied with definitions of technical progress but their taxonomic discussions should not be bypassed, because they bear upon and may even predetermine the views historians take about the mainsprings of innovation.

Technological progress is often defined to include a new device, process or good that makes it possible: (a) for firms and farms to produce a larger volume of output from a given bundle of inputs; (b) for firms and farms to supply qualitatively superior commodities and services from the same quantity of inputs; (c) for producers to offer new goods and services for sale directly to consumers. 'Novelties' are indeed an integral component of technical progress because markets can be saturated and economic growth (particularly in this century) has depended more and more upon persuading customers to buy unfamiliar products and services.

Economists distinguish 'techniques' drawn from the stock of technology available at any moment from new 'technologies' which add to available stock. Their models can cope readily with observed explanations for the choice of particular techniques by businessmen in market economies. For example, it is relatively simple to expose the rationale behind the American bias towards capital-intensive, and the Indian preference for labour-intensive, techniques of production. But models formulated by economic theorists are less adept in accounting for the rate at which new technologies accumulate through time.

Historians, preoccupied with long-term change, have devoted their own debates on technology to formulating distinctions between 'inventions' (seen as breakthroughs or discontinuities) and 'improvements' (defined as modifications to existing technologies and products). These distinctions are significant because they render explicit assumptions which often predetermine the views taken about the role of technology in the development of European economies. Thus historians who stress the continuous, cumulative and piecemeal nature of technical progress are unlikely to accord technology a primary role in modern economic history. Within their gradualist mode of writing, technical progress displays few discontinuities because it builds and improves upon the available and known stock of technology. For example, James Watt's separate condenser represents a mere modification to the steam engines of Savery and Newcomen, while the potentialities inherent in Watt's engine were developed further by Wolf and Trevithick. James Hargreaves's spinning jenny had its antecedents in Lewis Paul's carding machine. Precursors of Cartwright's famous power loom (brought anyway to a far higher level of productivity by Roberts) can be traced back to the ribbon looms and stocking frames of the sixteenth and seventeenth centuries. Finally, mineral fuels were used to smelt lead and copper long before Darby and Cort perfected their use in the manufacture of iron.

In this mode of historical writing about technology, the stress is invariably upon small sequential steps spread over long periods of time before full technological potential is achieved. As described the process is one of learning by doing and learning by using. There are feedbacks between users and makers of machinery and between consumers and producers of new commodities. Sometimes the realization of an innovation also waits upon developments in allied and complementary technologies. For example, steamships were not strictly speaking invented but depended upon the prior developments in metallurgy for boilers of sufficient strength to withstand the strains of steam-powered navigation. Similarly, durable steel rails and effective signalling and braking systems formed the preconditions for more rapid transit of goods and passengers by steam trains. By focusing upon the adaptation and diffusion of particular innovations over long spans of time, it is not difficult to show that reductions in costs which flowed from improvements and modifications to a prototype machine, process or commodity often exceeded by a large margin the decline in costs from the initial introduction of an innovation. Social returns imputed to an early blueprint or design can thus appear in the accounts to be rather insignificant.

Clearly definitions of technical progress that emphasize the continuous, gradual and cumulative nature of that process fit comfortably into the corpus of Marxist thought which sees technology as the inevitable outcome of the spread of capitalist commodity and factor markets across Europe. That same perspective is also congenial to neo-classical economists inclined to explain technological change endogenously, as a response to market forces and the profit maximizing drives of firms and entrepreneurs. Furthermore, if and when costs can be cut by technical improvements to installed machinery or if modifications to products increase sales then clearly market forces do set up pressures, demands and rewards for technological change. Gradualist interpretations also emerge in the writings of European historians inclined to see long-term advances in technology as the outcome of Western Christianity, which is said to have fostered a more manipulative view of nature than the dominant religions of other continents. Similar and often complementary interpretations depict technological change as a manifestation of Europe's evolving political and institutional arrangements, which allowed for greater degrees of freedom and individual enterprise than can be detected in the Ottoman, Mughal or Chinese Empires.

No doubt some nineteenth-century historians relied excessively on the 'heroic' theory of inventions, and the suppositions of modern social science tend to marginalize the role of individuals in explanations of economic development in general and technological change in particular. But historians of science and technology continue to write in terms of discontinuities. Perhaps the halls of fame are not crowded with men who can be considered as dispensable because sooner or later political, social or cultural forces would have created conditions in which individual discoveries became inevitable. On the contrary, the history of technology seems to be replete with 'breakthroughs' which had rather profound ramifications for the subsequent pace and direction of economic history.

By how long would the industrialization of Europe have been delayed without the prototype machines of Hargreaves, Cartwright and Watt? This is a pertinent objection to all gradualist and sociological interpretations of technical change. While they recognize that both improvements and modifications to technology in use can be modelled and explained, historians should continue to insist that the timing and location of key inventions were significant events in the economic development of Western Europe.

While such events may be unique enough to defy any attempt at general explanation, it still seems helpful to use the vocabulary of economics to separate forces promoting technological progress in national economies under those familiar headings of demand and supply. In brief, historians can organize the enquiry by posing two general questions: did technological advance in the century after 1750 tend to occur in those countries which had already built up the necessary capabilities to produce it more efficiently than other countries; or was it the case (as neo-classical theories of innovation predict) that pressures of consumer demand generated a response in the form of higher rates of technological progress in some societies (e.g. Britain) before and rather than in other societies (such as Tsarist Russia)? Both demand and resources are parts of any rounded explanation but the argument can be better structured under these headings. I begin with 'demand theories' because 'necessity as the mother of invention' is not only the layman's but also the economist's favourite explanation for technological advance.

Demand theories of technological progress

In plain language the demand theory is concerned to emphasize that technological progress satisfies wants. Without markets technologists simply produce curiosities of the kind exhibited in Leonardo's notebooks. Consumer demand is observed and transmitted to firms and farms in the form of rising prices for their products and is translated into projected profitability by businessmen and farmers. Price signals do not, however, automatically give rise to new technologies but are held to exercise a paramount influence on the rate and direction of innovation. In these models technology is almost subjugated to the pull of market demand. Economies experiencing an intensified pressure of demand, and which were incapable of responding elastically, were likely, within the constraints set by their available supplies of capital, labour and knowledge, to seek and thus to find solutions to supply bottlenecks in the form of new technologies. It is to demand forces that economists and many modern economic historians appeal when they attempt to explain the pace, pattern and location of innovations in the economic history of Western Europe during the industrial revolution.

Demand theories of invention, however, require rather precise specification if demand is not to be conflated with mere 'need'. At all times economies *need* technical progress, and whenever costs fall a greater volume of farm or industrial produce can be sold at lower prices. There was always some sort of expectation, at least in the economies of Western Europe, that markets could be created for new commodities and that markets already existed for cheaper consumer and producer goods. If they are to mean anything precise within the context of an attempt to account historically for the sources and location of innovations, demand explanations must imply that demand curves for particular outputs or inputs shifted outwards to the right. Thus if an historian enquires why a particular innovation emerged in, say, Britain in the 1770s, it will not suffice to observe that it represented a respone to changes in the pressure of effective demand. Why did the response occur then and not earlier when the pressure of demand might have been equally intense? Why did particular technological solutions or even clusters of innovations emerge in Britain rather than, say, in Holland, where demand conditions could have been equally propitious.

Furthermore, to establish any primacy for demand the historian must demonstrate that demand conditions, to firms and farms, changed prior to and more significantly than their costs of production. In theory it is possible to show that technological changes represented a response to rising prices for labour, capital and other inputs contingent upon shifts in consumer demand for particular commodities and services. But hard historical evidence for this kind of sequence, particularly for major European inventions in textiles, metallurgy and transportation, does not yet exist. Meanwhile there is no logical reason to claim that demand pressure acted as a dominating influence on technological progress. In a market economy any opportunity for profits would be taken regardless of whether it originated from a favourable shift in demand or a technical innovation promising high profits. Moreover, at the countrywide level of aggregation used in the writing of European economic history, it may well be impossible to estimate anything approximating to relative national differences in demand pressures. The normal historical assumption of a well perceived and growing market for the products of new technology becomes plausible if (and only if) historians can reveal the process of how signals from the market were transmitted and perceived by those engaged in producing machines, devices and new commodities. But when we turn to the case studies of European inventors at work from 1750 to 1850, it is not at all apparent that as a group they can be depicted as a body of men attuned to signals from the market. Jacquard, Arkwright and MacIntosh fit the description of profit-motivated inventors rather well. But Vaucanson, Berthollet LeBlanc, Watt, Cartwright, Crompton, Huntsman, de Chaurnes, Roberts, Fairbain, Kennedy and Trevithick appear to have been impelled more by scientific and technical curiosity than by any obvious drives to make money.

Finally there is an important question: if demand pull was the dominant influence on technological progress, how can historians explain the very long lags between the invention and the diffusion of Darby's coke smelting process, Kay's shuttle, Huntsman's crucible steel and Trevithick's high pressure engine – to take only a few random examples from British economic history? If a majority of innovations represented responses to the growing pressure of demand, diffusion, at least until 1914, seems to have been an inordinately protracted process even for the most commercialized economies in Western Europe. If, however, innovations were not widely diffused across industries and regions until the economic circumstances were ripe, the impulses behind any prototype invention remain hidden.

Demand theories of technological progress are difficult to validate or invalidate against historical evidence for the period from 1750 to 1850. Over that century markets did not transmit unequivocal profit signals to potential inventors – who in any case do not emerge from their biographies as a group of men particularly alert to such signals. As historians describe it, the whole process of invention seems shot through with uncertainties and chance. Recourse to demand theory except at the most obvious level of generalization (namely that markets for new technology had to exist) is not a particularly illuminating way to comprehend the origins of technical progress.

Supply or capability induced theories of technological progress

In order to generate a plan of blueprints for new technologies or consumer commodities, an economy requires skilled manpower, investible funds and a pool of usable scientific knowledge. Output (namely a flow of innovations) does not, however, vary in any systematic way with inputs (labour, capital and knowledge). There is no discernible production function for new technology but the economist's vocabulary of relating outcomes to inputs seems to be a useful way of structuring a discussion.

Between 1750 and 1850 technical breakthroughs and improvements were produced by scarce skilled and professional men who, as a group, almost certainly possessed common characteristics and motivations. Although abundant biographical information is certainly available at least for the major inventors of the period, that evidence has not been systematically studied and classified in order to offer historians a prosopography of these people. Such collective biographies have only been constructed for famous inventors of the twentieth century. But there would appear to be no *a priori* reason to suppose that contemporary evidence cannot be extrapolated backwards to apply, at least in general terms, to the previous century. Briefly stated, what we observe about major inventors active during the first half of the twentieth century is that: a majority were trained technologists; they made their mark at a relatively young age (thirty-five to forty years); they rarely admit to being motivated by prospects of pecuniary gain; and they had an uncritical attitude towards failure which meant they tried and tried again to find solutions to problems which they perceived largely in technical terms. This kind of group profile is not too dissimilar to the picture historians purvey, on the basis of more limited and unstructured evidence, of British and French inventors active in the century after 1750. A majority of inventors who grace the published literature for that period appear as technologists who were moderately well educated in the sciences (including mathematics). Few made fortunes and even smaller numbers are portrayed as relentless seekers after profits or even as men particularly well attuned to the dictats of markets. Many came from families lower down in the class structure but rarely from the bottom rungs of social ladders. In the United Kingdom a significant minority were non-conformists or Scottish. By the Victorian era these gifted amateurs of the early industrial revolution appear to have been giving place to a more professional and specialized group of inventors who took out several patents in the course of their careers.

Clearly societies well endowed with this specific sub-group within their work forces were more likely to generate innovations than countries less well endowed with such valuable human capital. But how particular countries and cultures came to possess an abundant supply of talented technologists at particular periods of history is very difficult to discover. National education systems, both private and public, clearly provided very different levels and types of training for their work forces. Before the late nineteenth century the utilitarian skills required to comprehend manufacturing processes and agrarian techniques were usually obtained on the job – which leads the search for the social and economic forces behind the formation of supplies of technologists into a circle. For example, from 1750 to 1850 most major breakthoughs in mechanical engineering occurred in Britain because by the mid-eighteenth century British industry already provided employment for a critical minimum supply of craftsmen capable of working with, modifying, improving and in the cases of the famous inventors, producing new prototypes which surpassed available machinery and equipment. This kind of explanation (paraphrased from several modern histories of technology) seems to take any enquiry way up the proverbial garden path.

Similar kinds of circularity and colligation problems afflict attempts to account for the rates, timing and location of technological progress in terms of an investment function. Perhaps British capitalists and landowners spent more upon the patronage of science and technology than their counterparts elsewhere in Europe. If correlations could be established between investment in discovery and improvements and the flow of innovations over time and across countries (a very tall order for historical research to follow) the correlations would still tell us little about the motivations, conditions and circumstances that promoted changes and variations in levels of investment devoted to the search for better technologies. They would merely form a preface to enquiry. Correlations define and never settle major historical questions.

The remaining input to discuss is science, which formed a pool of knowledge about the material universe and from which technologists could draw in their search for solutions to the problems of agricultural and industrial production. Such knowledge, embodied in books, journals and the persons of scientists, had increased in value and sophistication from the seventeenth century onwards. The connections between science and technology before the mid-nineteenth century are more difficult to discover than those of today, when the objectives of scientific research and the relevance of most scientific knowledge are more obviously utilitarian.

For earlier centuries, when the links between 'natural philosophy' and technological innovation were often not simple or direct, historians have certainly explored the whole range of possible connections. At one end of the spectrum science and the scientific revolution of the seventeenth century have been depicted as providing the basis for the technological breakthroughs behind the industrial revolution. At the other end of the spectrum are a larger group of historians who tend to see science and technology as rather distinct and even disconnected activities before the late nineteenth century. In their view the industrial revolution owed little to science. On the contrary science was largely concerned to classify and to explain the advances already made by craftsmen and farmers. For example, the theory of thermodynamics is said to owe far more to the development of the steam engine than the development of the steam engine owed to thermodynamics.

Predictably, the connections cannot be summarized simply and (chemicals apart) were neither direct nor apparent. In the course of the seventeenth century an interest in natural philosophy became part of the culture of educated men throughout Western Europe. Historians of science can trace 'advances' in knowledge in the fields of botany, zoology, pneumatics, ballistics, combustion and hydrology. But it seems impossible to trace links between these advances and the progress of agriculture or industrial production. Scientists in France

(Jars, Berthollet, Vandermande and Mange) also became actively engaged during the eighteenth century with gathering and classifying data on agricultural and manufacturing processes across a wide range of crafts and trades. They and their counterparts elsewhere in Europe published lucid memoires and encyclopaedias which appeared on the shelves of gentlemen's libraries or provided the basis for discussions at those literary and philosophic societies and academies which flourished in hundreds of towns throughout Europe (even in Tsarist Russia) during the eighteenth century. In these ways science emerged as a growing component of the culture and mentality of the educated classes. But the movement confronted tradition not merely as a challenge to existing social and political arrangements; scientific and experimental methods were also carried over into many areas of industry, transportation, commerce and even agriculture.

Biographical detail also reveals the extent to which a scientific education at a limited number of Scottish, Dutch, French and German universities helped to form many of the great engineers and chemists of the industrial revolution. Lower down the hierarchy of educational institutions historians have detected that mathematics, geometry, mechanics and machine drawing (studied at dissenting academies and other private schools) constituted an important component of the early education received by millwrights, wheelwrights, turners, mechanics, clockmakers and other artisans whose skills contributed so enormously to mechanical breakthroughs in industry and transportation.

Direct connections between science and technology are traceable for techniques and processes which involved some systematic knowledge of chemistry, particularly for the manufacture of alkalis, soaps, soda, bleaching agents and glass. Nevertheless, these connections and the role of scientific knowledge should not be exaggerated. In no sense can science be depicted as the propellant of economic development up to the mid-nineteenth century. Its connections with the accelerated growth experienced in metallurgy, engineering, textiles, brewing, transportation and agriculture are decidedly tenuous. Even for steam power the problems solved seem to have depended far more upon the contributions of craftsmen with skills in precision engineering and a practical knowledge of stresses endured by metals.

Too much has also been made of the experimental method as systematized and defined by academic science. Experiment and careful observation were hardly the province of a particular professional sub-group. Illiterate craftsmen and farmers had proceeded precisely along such lines long before the highly educated scientists began to observe, classify and reflect upon agriculture techniques and manufac-

turing processes in the seventeenth and eighteenth centuries. Furthermore, the observations and breakthroughs in science itself often depended upon effective collaboration with humble and unsung craftsmen who designed and constructed the optical and mechanical instruments necessary for numerous scientific experiments. For the industrial revolution, were the feedbacks from production to science not more significant than the forward linkages so carefully traced by historians of science?

Conclusions

Economic historians are particularly adept at importing theories and vocabulary drawn from economics into their accounts of the industrial revolution. They are agreed that technological progress, not only in industry but also in communications and agriculture, constitutes the *primum mobile* behind the accelerated rates of economic growth observed in Western Europe from 1750 onwards. When they return to explain the sequence, timing and location of innovations they find the economists' vocabulary helpful for the organization of historical discussion but are compelled to admit that these explanations are too general to be really illuminating. Thus there is a depressingly wide gap between macro theories of technological change and the micro-historical circumstances surrounding particular innovations. Until that gap is bridged the origins and sources of new technology remain obscure and the engine of the industrial revolution is appreciated but not understood.

Bibliography

C. Ballot, *Introduction du machinisme dans l'industrie Française* (Paris, 1923); M. Berg, *The Machinery Question* (Oxford, 1980); J. Daumas et al., *Histoire General des Techniques*, tome 11 (Paris, 1965); T.K. Derry and I.I. Williams, *A Short History of Technology* (Oxford, 1960); H.I. Dutton, *The Patent System and Inventive Activity during the Industrial Revolution 1750–1852* (Manchester, 1984); C. Gillispie, *Science and Polity in France at the End of the Ancien Regime* (Princeton, 1980); J. Kranzberg and C. Pursell, *Technology in Western Civilization* (New York, 1967); D. Landes, *The Unbound Prometheus* (Cambridge, 1969); P. Lebrun et al., *Essai sur le révolution industrielle en Belgique 1770–1847* (Brussels, 1981); S. Lilley, 'Technological Progress and the Industrial Revolution', in C. Cipolla (ed.), *Fontana Economic History of Europe, Vol. 3* (London, 1973); J.E. McClellan, *Science Re-organized* (New York, 1985);

S.T. MacCloy, *French Inventions of the Eighteenth Century* (Lexington, 1952); C. MacLeod, *Inventing the Industrial Revolution: the English Patent System 1660–1800* (Cambridge, 1989); P. Mathias, (ed.), *Science and Society 1600–1900* (Cambridge, 1972); *The Transformation of England* (London, 1979); A.E. Musson and E. Robinson, *Science and Technology in the Industrial Revolution* (Manchester, 1969); A.E. Musson (ed.), *Science, Technology and Economic Growth in England* (London, 1972); D. Roche, *Le Siècle des Lumières en Provence, 1680–1789*, 2 vols (Paris, 1978); N. Rosenberg (ed.), *The Economics of Technological Change* (London, 1971); *Inside the Black Box: Technology and Economics* (Cambridge, 1982); G. Rousseau and R. Porter (eds), *The Ferment of Knowledge: Studies in the Historiography of Eighteenth Century Science* (Cambridge, 1980); S.B. Saul (ed.), *Technological Change: the United States and Britain in the 19th Century* (London, 1970); J. Schmookler, *Invention and Economic Growth* (Cambridge, Mass., 1966); A.I. Usher, *A History of Technical Inventions* (Cambridge, Mass., 1954).

2
Resources and Technology

Peter Mathias

Resources and economic development

It is unfashionable now in development economics and economic history to give much prominence to the natural resources of a country when assessing the reasons which have favoured or retarded its industrialization. Resources (collectively or individually) do not feature in S. Kuznet's comparative survey *Modern Economic Growth*; the chapter on 'Population and resources' in A. Lewis's *Theory of Economic Growth* is almost entirely devoted to a discussion of population and relative scarcities (particularly agricultural scarcities); Meier and Baldwin allow resources scarcely more than a page in *Economic Development* – and that primarily concerned with resource-depletion in advanced economies.[1] Population, capital, entrepreneurship, foreign trade, economic policy, motivational structures and cultural norms all seem to have much more space devoted to them. At the end of the nineteenth century and beginning of the twentieth, in contrast, priority of importance was given to the abundance and cheapness of coal in the UK and its scarcity and expense in France as a principal explanation of the precocity and extent of industrialization in England compared with France. Alfred Marshall gave great weight to this, as did Sir John Clapham and contemporaries to the process, whether complaining of French deficiencies like Chaptal and Gabriel

---

[1] S. Kuznets, *Modern Economic Growth: Rate, Structure and Spread* (New Haven, Conn., 1966); A. Lewis, *Theory of Economic Growth* (London, 1955); G.M. Meier and Robert E. Baldwin, *Economic Development: Theory, History, Policy* (New York, 1957).

Jars, or worrying about the depletion of coal reserves in England like Jevons, who proclaimed in 1865: 'Coal stands above, not alongside, all other resources'.[2] Natural resources have more recently been stressed by E.A. Wrigley. Not incidentally, perhaps, he was a geographer by training.

I do not think this is accidental and there may be justification for it in discussions of the relative importance of factors in economic development in a twentieth-century context. But resource patterns are very easily taken for granted, and to project this pattern of priorities back into the analysis of the factors favouring (or inhibiting) industrial growth in the late eighteenth and early nineteenth centuries is seriously misleading.

A favourable resource position was more important, I would argue, in early phases of industrialization, in the historical conditions prevailing up to the second half of the nineteenth century; it has become steadily downgraded since then, particularly for the sophisticated industrial economies. In the late twentieth century it is remarkable what natural resources a successfully industrializing economy can do without, and the more successful an industrial economy becomes the more quickly, in a context of rapid rates of growth, will it be likely to outrun its local resource base. Japan is a case in point now, as was the United Kingdom during the nineteenth century. Some of the most successful industrial economies recently have been born virtually without a resource base at all. Hong Kong, for example, does not even have sufficient local water: normally adequate water supplies and an equitable climate remain very important resource attributes for an economy, even when most of its other raw material resources for energy, industrial processing, or even foodstuffs may be imported. The reasons for this latter-day downgrading of the importance of a favourable local natural resource endowment are clear. Applied science is creating a steadily lengthening list of synthetics, which discounts the availability of local natural resources. Many of the manufactured products and the exports of sophisticated economies tend to have a raw material content which forms a very small percentage of their total price, which means that the transport costs of those raw materials are not very significant in relation to their total costs. Energy imports (particularly since 1973) form a partial exception to this generalization. Above all, the fall in bulk transport costs in a world of supertankers, container ships, very large bulk carriers, trains and trucks means that

---

[2] A. Marshall, *Industry and Trade* (London, 1923;) J.H. Clapham, *The Economic Development of France and Germany, 1815–1914* (Cambridge, 1921).

there is no prohibitive economic cost, at least in times of normal international relationships, in bringing most of one's energy requirements, basic foodstuffs and industrial raw materials from the other side of the world.

The position was very different before the development of railway, canals and large iron steamships, when overland transport costs, and physical criteria, were limited to the economics of the pack-horse and wagon. This is more particularly the case where economic activity, and manufacturing in particular, was dependent on cheap bulk materials (for energy inputs above all) in industries which were great absorbers of energy and of raw materials that lost great bulk in processing. The manufacture of metals was the clearest example. Very high overland transport costs precluded a mass production, low-cost metal industry without immediately local raw materials: essentially cheap mineral fuel. The logistics of transport costs and energy inputs (it took up to 10 tons of fuel to make a ton of refined iron and up to 20 tons of fuel to make a ton of refined copper in the eighteenth century), coupled with the technical aspects of materials, precluded areas where iron ore was separated by long distances from charcoal or mineral fuel from becoming a base for heavy industry. For example, the area of woodland from which charcoal had to be drawn to serve a blast furnace, the cost and physical difficulties of carting charcoal any distance without crushing it to powder, and the behaviour of charcoal in blast furnaces (where it compacts under great weight) all combined to enforce an iron industry of widely scattered, very small units. Expansion with such a matrix of raw materials and technology enforced a greater diaspora of tiny blast furnaces and forges to more remote woods and streams, in a search for sources of charcoal and power for the tilt hammers. This precluded concentration at sites favourable for the market or the labour force, larger plant and furnaces, economies of scale and external economies leading collectively to low-cost mass outputs. Similarly, with the economies of horse-and-dray transport, the only context which could lead to the evolution of the large brewery with massive economies of scale was the intensive immediately local market of a great city such as London. Beer would double its price in a journey of over ten miles by dray; so would timber and other cheap bulk staples.

This is not so clearly true of the textile industry or small metal wares, but it became true whenever the scale of output grew and also where technological change led towards greater energy consumption. In the cotton industry, for example, the raw material for all European economies came from far distant lands, by sea. There was a 'magnetic pull' from overland transport costs to keep processing centres close to the ports, and supporting ancillary industries, such as dyeing, bleaching and calendering, were becoming fuel-intensive processes by the end of the eighteenth century. Bulk production of simply processed materials combined with productivity advances in processing means that the proportion of raw material costs in total unit costs rises. This is particularly true of an industry where great economies of scale, particularly in labour costs, were available – as with brewing and distilling, or later with other food-processing industries.

Considering the role and importance of resources as an economic input, one cannot take them as an autonomous, absolute 'given', a sort of 'unmoved mover' in the process of growth. Their valorization – the economic significance of their existence within an economy – is a function of transport costs in particular, of the technological matrix current at the time, and of the level and trends of demand – to name only the most important of these contextual considerations.

Resources and technical change

The relationship between resources and technology represents a two-way reaction, to a degree. Technology itself adapts and grows in the direction of available resources, or the cheapest resources. As incentives for substitution are created, by a rise in price or progressive scarcity of resources used in the traditional 'mix' of production, so technical change will tend to move in the direction of other resources, if these possess economic advantages. The technological base will also change, and new resources become valorized, from the harvest of advances in scientific or other sorts of knowledge, from the yield of curiosity, even without such direct economic incentives. But commercial profitability, dependent upon relative factor prices, will at least govern the 'mix' of inputs and resources for existing products where substitution is technically possible. The expansion of output with unchanging resource inputs and static technology will at some point lead to a breakdown in the equilibrium between resources, technology and the prices of inputs. This at once creates incentives for substitution – for adapting other resources, at lower costs, and employing new technology to do it. Technical change can get built into this sequence with sufficient commercial responsiveness. For example, the move to smelting iron with mineral fuel in England, introduced in 1709, did not spread until the 1760s, because there were few incentives for it until the progressive increase in costs from traditional materials valorized 'coke-smelting' at that point. Wilkinson's introduction of coke-smelting to France in the 1780s was a failure, as were attempts in Russia, for the same reason. Not until the 1850s in these countries did the relative costs of coal and charcoal as fuel tip in favour of coal and lead to the rapid spread of mineral fuel in the iron industry of Northern France.

There are important subordinate linkages and paradoxes about a natural resource endowment – but, again, they have to be seen not autonomously but in relation to the intensity of demand (and hence relative price movements) focused upon the conjuncture of resources, technology, markets and potential technical change widening the alternatives of the resource-technology mix available. With the basic resource inputs for medieval technology, Britain was fortunate in her progressive deficiencies in two important matters. There were shortages of wood for fuel (particularly in London), in the domestic market and for industries such as brewing or dyeing, and above all for charcoal for iron-making and other smelting. Water power was also short, particularly in the Midlands, because all available mill-sites were occupied. This meant that incremental demands in industrializing and urbanizing localities were driving up prices of wood and water power relative to alternatives, and sharply increasing the incentives for adopting, and discovering, substitute resources by technical change. Given the double fact that the alternative technology was that of mineral fuel and iron, allied with engineering skills, and that the resource position of Britain was peculiarly favourable to the new matrix, this deficiency in traditional resources proved highly favourable to industrial advance, providing both a 'push' and a 'pull' impetus. The absence of *either* constituent would have increased the inertia exerted against change: running short of timber and water power in the absence of cheap coal and iron would merely have pushed up prices, increased imports of traditional raw materials, raised the costs of economic expansion absolutely and disadvantaged the British economy relative to others more bountifully endowed with traditional raw materials. In the event that bounty in continental Europe and North America retarded in those areas the transformation of the metal industries, the growth of their coal industries and the whole complex of mineral-fuel technology, power technology and engineering skills fused into the new matrix. Not accidentally the initial leadership by the United States in certain engineering technology in the nineteenth century was based upon woodworking equipment: a technology which was wasteful in its use of material but economized on skilled labour.

Britain gained greatly in all fuel-intensive industries because of the abundant, easily won coal which was widespread throughout the industrial regions (and indeed conditioned the existence of many of them). Coal could be used virtually at the pit-head in many localities (and actually occurred in adjacent strata with 'black-band' iron ore in Scotland) or with haulage over short distances. Transport costs were the key to the price of coal at processing sites. Cheap fuel, cheap energy, cheap iron conditioned the path of innovation in Britain relative to France. Technology change developed along more fuel-intensive lines in Britain, with a more prodigal use of coal, for current energy use and with capital equipment which incorporated high energy inputs in its manufacture. Energy-intensive operations were the basis of much industrial growth in this period in the heavy industry and construction sectors.

During the eighteenth and early nineteenth centuries Britain was unique in the extent of its use of coal; and was recognized as such by interested visitors from other countries, who drew the contrasts with their own deficiencies in this respect. In fact this had characterized the British economy since the sixteenth century. Urban, fuel-intensive industries in London and other principal towns on the coastline – brewing, dyeing, soap-boiling – as well as domestic hearths were relying on 'sea-coal' and smoke pollution from coal-burning was already a problem for the metropolis. During the seventeenth century other fuel-intensive industries in England and Lowland Scotland had moved over to a mineral-fuel base: salt-making (from brine), brick-making (where the cost of fuel was several times greater than the clay), malt-kilning and the smelting of all metals except iron and steel (by 1720). Blacksmiths and foundrymen also used mineral fuel. The same was true of the glass and pottery industries, much of the chemical industry in the eighteenth century, where salt and soda-ash were basic raw materials, and the finishing processes of the textile industry (calendering, bleaching, dyeing). When the smelting and refining technology of iron became based on mineral fuel during the eighteenth century the metal industries became a major market for the expanding coal industry, which determined the future location of the furnaces. Coal, in that century, became the basic domestic fuel, not just in the towns but over the whole of the coastal belt, penetrating inland via the navigable rivers and canals by the end of the eighteenth century before distribution was universalized by the railways after 1830.

Coal output in Britain was already about 3 million tons per annum when the eighteenth century opened, and had grown to 15 million tons at the century's end. This was an unprecedented volume of output, unmatched by any other traded commodity, with major implications for the logistics of its transportation. More than a third of the merchant fleet plying coastwise carried coal from the North-east field (predominantly to London) and most canals in the second half of the century had coal as their main freight. Mining (for tin and copper ore as well as coal) was also the main stimulus for the emergence of steam power during the eighteenth century. Coal output (almost all of it for use within the country) was growing at over 2 per cent per annum compound in the second half of the eighteenth century, faster than the economy as a whole. Energy consumption doubled per head of the population over the century and energy inputs (not all coal) probably increased by 50 per cent per unit of total output. Britain was becoming a more energy-intensive and fuel-intensive economy. Perhaps half of total output was taken by the domestic market in 1800, with a third, in all, for direct industrial use (10–15 per cent by the iron industry alone).

Coal prices broadly moved in step with general price levels, as a national average, over the century. Diminishing returns did not set in, despite the sustained expansion of output: bountiful resource endowment in many regions, cheapening transport and improved technology (particularly techniques of drainage and ventilation in the eighteenth century) offset the cumulative problems of attrition in an extractive industry. But if coal prices did not fall dramatically in London they evened out (and fell dramatically in the process) as the marketing range of water-carriage spread. When moved by horse-and-cart, coal prices doubled every few miles. Retail prices of coal halved in Manchester when the Bridgewater Canal was extended to the city. And, more important for many industries, much coal could be used very close to the pit-head, because of the extensive distribution of the coal measures in England, South Wales and Scotland. Average national prices, based on institutional prices in the South and in London, therefore do not at all reflect the level of costs of coal in the main industrializing regions.

At a later stage in the nineteenth century that paradox was repeated on the other side. Once, belatedly, the French iron industry was pushed into the new matrix, with coal much more expensive than in Britain the increased costs induced important technical innovations in fuel economy in blast furnaces, which led forward into research and more advanced techniques. Low-cost mineral fuel in Great Britain encouraged wasteful techniques and discouraged research and applied science in metallurgy in the British iron and steel industry. French locomotives were more costly than British: they were more sophisticated and designed for greater fuel economy. The world economy has seen a similar sequence more recently with cheap energy relative to other input costs in 1945–73 encouraging energy-intensive technology; this was then seen as a very high cost technical matrix post-1973 with relatively high energy price in the pattern of factor costs.

Another very important subordinate linkage to resources is that of skills. Certain skills, more particularly before the late nineteenth century, tended to be localized on a resource base. Techniques for diffusing skills remained very empirical (always involving physical migration of skilled workers and artisans). The techniques themselves were usually acquired by 'learning by doing' rather than by formal knowledge, which was often still unknown, and these techniques became very specific to local materials: the design of the blast furnace, the technique of loading it and the regime of firing it were all critically dependent upon the exact nature of the local coals and ores. These materials are not, as they so often appear in textbooks of economics and economic history, homogeneous commodities. In reality they exist in bewildering varieties, with widely different 'mixes' of properties. The development of skills associated with the new resource base of mineral fuel, ranging from engineering skills to precision large-scale metal working, had crucially important linkages and spread effects in the economy generally. Even now, in the very different world of the mid-twentieth century, we can see the importance of skills being localized as a resource base in the initial absence of British firms and technology able to take advantage of the opportunities offered by North Sea oil exploration. The only large immediately available pool of such technology, know-how and men was the USA: predictably Aberdeen was full of Texans.

Transport costs, and hence the location of natural resources, are also critically important to their valorization as economic inputs, rather than as geological phenomena.

Innovation and technical change

Conditions of innovation and changes in technique are virtually never simply an individualist or 'heroic' process. Of course, all innovations are ultimately consequential upon the endeavours and actions of individuals, in common with all other aspects of societal and human change (such as population growth). But, in the early stages of industrialization at any rate, they are more to be conceived of, analytically, in terms of biological evolution. One must distinguish between accretions to the stock of knowledge and invention, innovation and the diffusion of innovations.

Formal knowledge advances under the stimulus of intellectual curiosity as well as more material incentives. Even though inventions imply a search to apply knowledge to techniques, and are commonly much influenced by the hope of personal reward (which means the successful attainment of utility somewhere), the record of inventions is not governed nearly as much as the process of *innovation and diffusion* by commercial criteria. However, inventions will only be put to productive use, and taken up more widely in a market context, if they offer commercial advantage compared with the existing techniques. This may come from a relative saving of costs from labour, capital or resources (or from more than one of these factors). Such costs have to be broadly conceived – information about new techniques may be difficult to acquire, new skills associated with them long in learning, or expensive to hire. In special fields of endeavour prices may not be so critical. Where the ruler or the state is the purchaser, or when inventions are destined for a luxury market, costs may not be a determining consideration (as they never have been, for example, in military technology, where functional efficiency and technical superiority have been more persuasive arguments). Where innovations have occasioned new products, detailed cost comparisons with older technologies producing non-competitive products cannot be applied so exactly: qualitative differences always complicate the measurement of comparative advantages. But, with such reservations, most innovations in Western countries governing the level of productive techniques in these economies during the eighteenth century made their way in a market context, and commercial criteria therefore set the measure of incentives for their diffusion.

Technical change – the flow of applied inventions – is not analysable simply as the autonomous arrival of new techniques from a fund of knowledge built up, like a library, exogenously to the productive context. The 'evolution' of innovations comes from an accretion of changes resulting from very widely diffused endeavour, a very high proportion of which never comes to fruition. Only the advancing edge of technical change appears above the surface of history. A very small percentage of patents, for example, were actually put into practice (and very many fewer earned their costs in patent fees). In many fields, such as ship design, hand-tool design and seed selection in agriculture before the very modern period, there is a close analogy to natural selection techniques, in that improvement came from widespread empirical endeavour, widely diffused, gradual, 'blind' in the sense of not being the self-conscious product of a single will or identifiable set of wills. The fittest survived by the tests of technical and commercial viability, and gradually drove out the less efficient by similar sanctions in a sort of technological and commercial Darwinism.

A long lag commonly existed between invention and adoption, between invention and innovation, between successful innovation and widespread diffusion. Indeed access to formal knowledge or laboratory demonstrations of technique were usually not the key instigators to technical change, particularly as far as the timing of the diffusion of 'best practice' technology is concerned. The 'scientific' or 'formal' awareness of many key innovations long antedated their adoption and diffusion. In other cases equally long time lags existed between the spread of an empirically discovered new technique and the formal awareness of the relationships involved in it. The time-scale of the impact of such additions to formal knowledge in economic history follows very different criteria to the acquisition of such knowledge (or even invention). In the eighteenth century, as evidence, almost all the changes in agricultural techniques in new plants, crop rotations and animal breeding, and some new implements, can be traced back to seventeenth-century or earlier sources. Rotative motion and steam power, separately and in conjunction, were known about, in the formal sense, long before 1781 (the date of the Watt patent). The essentials of mechanized cotton spinning and power-weaving techniques were long known, and in the case of cotton spinning the elements of water-frame technology were actually innovated by Lewis Paul in the 1740s. Power weaving spread over the textile industry very slowly, during fifty years or more. The only important technical innovation in the transport investment of canals which had the major impact on industrial costs in the eighteenth century, was the pound lock, which had been known since the fifteenth century.

Most of the innovations associated with the early decades of the industrial revolution period were, in fact, basically medieval technology pushed to its limits – new combinations of known devices in textiles, mill-technology with water as the prime source of 'massive' power and means of transport. Steam power would be an exception, but formal knowledge about the 'power' of steam long antedated this innovation and steam power, though a crucial new prime mover at the frontier of technical advance, activated only a tiny fraction of total rotative motion in the eighteenth century. The same is true of certain central pieces of technology in machine-making and machine tools – such as the screw-cutting lathe and slide-rest, which had been known in arsenals, scientific instrument-making and the watch and clock industry long before the emergence of engineering workshops. Lags between innovation and diffusion – the gap between 'best-practice techniques' and the general level of technique – tended to be shorter following the demonstration of practicability, but diffusion of techniques was still astonishingly slow judging by later experience. Agriculture was particularly slow to respond, for reasons which reveal much about the process of technical change in the eighteenth century.

Further phenomena in the process of technical change in this period were the extent of simultaneous, spontaneous inventing and the extent of 'substitute' inventions, as people sought different ways of performing a function, solving a problem or bypassing a blockage. All these phenomena show how technical change was a response to the demands of the context.

Each sequence of innovation can be broken down into some discrete changes with identifiable authorship, but these rest upon 'continuum' improvements, accreting gradually and anonymously, which integrate the evolution of the innovation with its context. This is partly a question of 'on the job' adaptations of techniques; partly because many innovations depend, for their own effectiveness, upon correlated, but distinct, advances in other techniques – just as new spinning techniques depended for their viability on the spread of iron machinery, or steam-engine efficiency on rising standards of accuracy in working iron. Technological advance at the frontiers of technique also shows a high degree of inter-relatedness. A new device may be frustrated, not because of any fault in the concept or any lack of formal knowledge, but simply because materials do not exist of sufficient strength to make it possible, or prime movers of a sufficient power–weight ratio to make it move. More efficient engines or stronger metals may therefore valorize a whole backlog of technical change awaiting embodiment. This is characteristic of large numbers of Leonardo da Vinci's inventions, thrown off from a brilliant imagination long before there were technical means to realize them. In a similar way, advances in present-day aeronautical engineering are often dependent upon innovations in metallurgy. The missing elements in complementarity are not only technological, of course: possibly there is not sufficient demand for the product, or transport costs may be so high that effective marketing on the minimum necessary scale is impossible.

Apart from such technical interdependence, innovation and the widespread adoption of new techniques were also responsive as much to the economic context as to the 'commercial Darwinism' which winnowed the 'technological Darwinism'. Putting inventions to productive use involved all the costs and problems of translation from laboratory technique to industrial production, from the largely non-commercial context and movitations of the pursuit of knowledge to profitability as a condition of existence. The determinants of timing for innovation and diffusion were usually set by non-technical criteria, particularly being a function of demand of market criteria in different ways, whereas accessions to knowledge, the advance of science and the record of inventions were not governed by this in the same direct sense. The expansion of output, even the absolute size of an industry, could prove a stimulus to cost-reducing and quality-improving advance under competitive conditions. Adam Smith's principle of the division of labour being limited by the extent of the market can be supplemented by another proposition that 'inducements to innovate' are limited by the extent of the market – a 'demand-pull' model, in fact. Research, design and development expenditures, for example, can be carried more efficiently as their costs are lowered per unit of output. Extensive machinery and large-plant operations increase 'fixed' or overhead costs, which create incentives for keeping plant running to capacity, and this can lead to the need to extend markets and cut prices to sell more. This sequence sets up further pressures to seek innovations and cut costs more. In a competitive situation, in the absence of effective combinations to cut output, this proved a self-reinforcing sequence, given the conditions that elasticity of demand kept up and that opportunities for further innovation remained.

In a competitive context, provided that available innovations offer reductions in costs for making existing products, or the opportunity of putting attractive new products on the market, technical change is likely to proceed, even without any changes in relative prices. But where the relative prices of materials, capital or labour embodied in a product do change then a further inducement for innovation can be created. An increase in the demand for a product to which an advance in technique is applicable, or of its raw materials or components, or collateral industrial products, is likely to become expressed in a growing shortage of a factor of production, a scarce skill or some bottleneck or other which is reflected in a change of factor prices. The price of the scarce factor is driven up, upsetting the previous price equilibrium and occasioning incentives for changing the balance of inputs by way of some change in technique.

We should also ask whether military sponsorship was productive of important innovations in the late eighteenth century, bearing in mind the very high rate of naval and military spending at the time (compared with investment flows in industry). In the twentieth century wartime emergencies and peacetime defence spending have proved the forcing ground for key innovations and development laying the basis for a whole range of industries which were major growth sectors of the second half of the century, atomic energy, computers, jet-engines, electronics, aeronautics and metallurgy among them. Such 'public sector' spending on research and development for military technology in the eighteenth century, as today, did not have profitability as a condition of its existence in the same way as spending to promote innovation in the private sector. Yet in the eighteenth century new swords did not produce such important ploughshares. Nor were the swords themselves strikingly new. What it is now fashionable to call 'spin-off' from military spending was small and was to remain so, broadly speaking, until the second half of the nineteenth century.

In the industrial revolution period a large number of innovations of the second order of importance can be identified. The most well-known is John Wilkinson's cannon-boring device, patented in 1774, which enabled him to produce a true cylinder for steam engines. From Admiralty commissions came improved cartography, exploration and surveying, some improvement in the design of ships, the use of iron in subsidiary parts of ship construction, the copper sheeting of ships, the famous chronometers (developed in connection with the struggle to find a more accurate means of establishing longitude at sea), certain medical and dietary innovations, preventive medicine and the like. Other examples can be quoted but they are at a similar level of importance. Naval and military arsenals had developed the most accurate standards of metal working, and employed the best machine tools available. Because so much weaponry in Britain was put out to private contractors, capacity and skills in the primary iron industry (among the casting shops) and in the Birmingham trades owed something to military funding, particularly in wartime.

It can be argued that there is, in a broad sense, an 'ecological balance', even if never absolute, between the scale of production, a particular 'mix' of raw materials, the technological matrix (and its embodied skills), forms of power, modes of transport and even certain institutional forms and business structures, which find themselves in equilibrium. This has doubtless never been entirely without strain and pressures, and such pressures have been accommodated through technical change altering the proportions of input and some substitution of factors – even in medieval times. But the rate of change and the build-up of disequilibrating forces was then very slow. During the eighteenth century this pace quickened and the technological disequilibria became a forcing house for further innovation.

The particular inputs whose growing scarcity or rising costs – or absolute level of costs – stood to exert the most widespread leverage for inducing changes in technique were probably labour, water power and fuel. Wage costs being such a high percentage of total costs in most artisan or 'domestic' industry, the scope for economizing by mechanization was very large. In this respect the pressures for innovation to economize on wage costs – particularly the wage costs of skilled artisans – were increased by the rise in the level of wage rates in the economy between 1750 and the 1780s. The lag in the diffusion of smelting iron by coal until the 1760s, mentioned above, was probably not significantly influenced by secrecy (as many early commentators claimed) but by the general absence of expansion in the industry, which did not put increasing pressure on charcoal supplies, coupled with extensive imports of bar iron from Sweden and Russia. When output expanded after 1750 prices of charcoal were driven up, creating a major incentive for adopting mineral fuel in smelting. Increasing demand also means the need for a decision to increase capacity in plant and machines. There is a higher propensity for incorporating technical change when such decisions have to be taken on expanding capacity than if existing – and still working – plant has to be ripped out and newer machinery substituted. An economic context of expansion, that is to say, is more favourable to technical change than an economic context of stagnation – for this as for other reasons.

Once an economy is on the move, with innovations flowing and the scale of output in various industries rising, the process can become self-sustaining – or at least powerful incentives for this become built into the sequence. Even a depression in this context, by putting pressure on profit margins, can create incentives for innovations to cut costs, particularly innovations not requiring heavy investment. One innovation breaks an equilibrium in a traditional sequence of processes, creating a distortion with others, which demands further innovations to seek a new equilibrium: the great difference in levels of productivity between the flying shuttle and the spinners, in a famous instance of this, created great incentives for improvements in spinning techniques, leading directly and self-consciously to the development of the jenny. Factory spinning created similar incentives for power weaving. The distortion or disequilibrium could be one of materials as well as of the flow of production between different processes. When a steam engine was hitched to wooden machinery it shook it to pieces and required the innovation of iron machinery. That innovation allowed more complicated, heavier machinery, which itself demanded more powerful prime movers. This lay directly behind Watt's development of the double-acting engine.

These disequilibria operated between manufacturing processes and the forms of power needed to sustain them; between production and the forms of transport needed to clear the goods and supply the raw materials – particularly the bulkiest of all, coal for energy inputs. It operated between one industry and its subsidiary industries. Before the railways, the great and rising demand of the cotton spinners was the greatest single inducement to the specializing out of an engineering and machine-making industry quickly supported by a machine-tool industry, with a wholly new range of techniques growing up in the shadow of the mills. Sometimes the disequilibrium was one of time. Traditional techniques of bleaching, a cottage affair using the sun and sour milk as its main agents, taking up a very large amount of ground and requiring some months to complete the process cycle, were incompatible with the great increase in yarn and cloth as productivity rose in the main branches of the textile industry. Rapid 'chemical' bleaching with chlorine became imperative in such circumstances. The same was true of methods of cotton printing: the demands for a speedier technique than hand-block printing produced the rotary method. The disequilibrium might be one of speed, of scale of production, of power, of efficiency, of materials; they would all be expressed in an increase in costs where the disequilibria impinged on the sequence of processes.

In circumstances where techniques are fairly stable, when demand is fairly constant (or falling) and the general momentum of change in an economy low, such technical 'frictions' which imply economic costs are likely to bring the original innovation to a halt. Disequilibria can then act as brakes on innovation, and the circle becomes a vicious one. When the general momentum for change is high, when an economy is on the move with self-sustained growth in progress, then friction can invoke the response of further innovation, creating a chain reaction which converts the whole sequence, all the linkages, to the new norms – the new rhythms, the new tolerances, the new speeds, the new materials, the new prime movers.

The last twenty years of the eighteenth century saw the leading edge of technical change in strategic industries experiencing such a self-sustaining sequence, which led back again and again to coal, iron, machinery, engineering skills and steam power, and emphasized the advantages of location near cheap coal. The beginnings of this new logic were becoming established – only the initial stages of such a logic, affecting a small arc of innovation, yet the strategic leading edge of technical change none the less. Disequilibria were continually being resolved by innovations which increased the reliance on this new matrix of industries, materials and skills. And it is exactly in the combination of accessible, cheap coal and iron ore – the strategic materials – that Britain's natural resource position was ideal. And by

1800 it was exactly in the skills associated with the strategic new industries of iron, mass-production iron-making, metal manufacture, large scale metal-working techniques and engineering (particularly mechanical engineering and power engineering) that her lead over continental economies was most marked.

Science and technology during the industrial revolution

A long-standing debate, which still continues among both historians of science and economic historians, concerns the relations between the advance of science and technical change in the eighteenth and nineteenth centuries. W.W. Rostow, T.S. Ashton, J.D. Bernal, A.E. Musson and E. Robinson are among those who have stressed the links between science and innovation in the industrial revolution. A.R. Hall, Sir John Hicks, David Landes and Eric Ashly have argued that the links did not become significant or sustained until after 1850.

It is easy to point to a long list of piecemeal examples of a positive association between science and innovation, at least in broad terms. Amateur scientists were certainly prominent in branches of the chemical industry, with bleaching, chlorine, sulphuric acid, coal-distillation and soda-making as examples. Dr Roebuck, Charles and George Mackintosh and Lord Dundonald were prominent as entrepreneurs in the chemical industry. What of James Watt and steam power? Further indirect support can be claimed from the surge of interest in science and 'improvement' in the eighteenth century. The Lunar Society of Birmingham, established informally c.1765, was the exemplar for a large number of 'philosophical' societies, established in such small towns as Spalding and Northampton to pursue amateur interests in all manner of 'natural knowledge'. The Manchester Literary and Philosophical Society (1781) became the most famous of these provincial societies, but they all lived in the shadow of the Royal Society of London. The Royal Society's original statutes declared an intention of relating science to improved practice no less than the later metropolitan societies established to promote this end, such as the Society for the Encouragement of Arts, Manufactures and Commerce (1754) and the Royal Institutions (1799). The preamble (of 1662) ran: 'The business of the Royal Society is: to improve the knowledge of natural things, and all useful arts, Manufactures, Mechanick practices, Engynes and inventions by experiment.'

The evidence of *intention* and of the endeavour to promote science for utilitarian purposes is pervasive. It forms a continuum from the

early seventeenth century (and even, it can be argued, from as far back as Roger Bacon in the thirteenth century). Scientists, industrialists, spokesmen for governments and publicists agreed in chorus that science should be promoted (and funded) for practical purposes. The title of Robert Boyle's survey of industrial methods, listed with a view to their improvement by science, sets the canon: *The Usefulness of Natural Philosophy* (1664). A technical treatise on brewing, written in 1804, remarked: 'Chemistry is as much the basis of arts and manufactures, as mathematics is the fundamental principle of mechanics'.[3] These assertions run through until our own day. They were supported by a stream of official endeavour, which brought some success: Admiralty-sponsored research into cartography, astronomy and exact time-keeping (to improve methods of discovering the longitude of ships at sea), medical research into scurvy, concern with ballistics and the like. Many scientists concerned themselves with industrial processes and, as T.S. Ashton put it, 'there was much coming and going between the laboratory and the workshop' in eighteenth-century England and Scotland.[4]

All this is not in doubt. However, a critical assessment of the claim that scientific advance was a (or the) prime determinant of technical change in this period has to be made. It is generally agreed that the links between scientific advance and industrial progress became much closer after 1850, and were consolidated by new institutional connections. This should modify the extent of the claims being made for the connection in the eighteenth century. The evidence of result is not nearly as convincing as the evidence of intentions, when analysed *ex post facto* with hindsight. Scientific knowledge was growing internationally, not in national isolation. The scientific movement was common to countries across Europe (the Academy of Sciences in St Petersburg was founded in 1724). Science had become formally institutionalized more strongly in France than in Britain. All this can be summarized in the generalization that the accumulating stock of knowledge (the 'science' variable) does not show nearly the same national variations as does the speed and extent of innovation in different European economies.

All the problems already discussed show how important other determinants of timing were compared with new acquisitions to the stock of knowledge. However, this is not just a question of the advance of science being a necessary, but not a sufficient, condition of technical

---

[3] R. Shannon, *Practical Treatise on Brewing* (London, 1804).
[4] T.S. Ashton, *The Industrial Revolution, 1760–1830* (Oxford, 1948), p. 16.

change. In fact, much too simple assumptions have dominated the debate about the relationships between scientific knowledge and technical change in the industrial revolution – particularly that the nexus was one-to-one, direct, linear and a one-way street where the direction of the flow of traffic was exclusively from science to industry. Accumulating a piecemeal list of instances, where the connection was positive, or where at least there was a connection between science and industry, does not, of itself, address the question of how representative such instances were of the general process of technical change. It also begs the question of how far the impetus deriving from industry was able to produce the conditions of innovation needed to sustain its own progress – coming from within its own empirical world and not given to it from an 'exogenous' world of science advancing under its own complex of stimuli. That in turn begs the question of how far industrial demand, with the needs of the 'empirical' world, was itself the stimulus for creating new scientific knowledge in this period, of what 'feeds-back' or 'feeds-forward' there were, what indirect as well as direct linkages.

In my view the empirical stimulus creating response within the immediate context of production was the dominant theme and accounts for a very high proportion of the advance in productivity, and was the greatest determinant of the timing and rate of diffusion of new techniques, even in those industries most exposed to the impact of science. Great areas of advance remained relatively untouched by formal scientific knowledge, judging by results rather than by aspiration or endeavour, until the nineteenth century – agriculture, canals, machine-making, iron- and steel-making, the mechanization of cloth-making. A very small proportion indeed of the labour force was engaged in trades where the linkages were – superficially at any rate – high, as with the chemical industry and steam power. Nor, by and large, were innovations a product of the formal educational system of the country. It was more a question of great determination, intense curiosity, quick wits and clever fingers, getting a backer to survive the expensive period of experimenting, testing and improving, which all tended to be more important than a scientific education. Most innovations were the product of inspired amateurs or brilliant artisans trained as clockmakers, millwrights, blacksmiths, carpenters, or in the Birmingham trades. William Murdoch, the chief mechanic of Boulton and Watt's engine manufacturing, was more representative than James Watt. They were mainly local men, empirically trained, with local horizons, often very interested in things scientific, aware men – but men responding directly to a particular problem. The chemistry of iron making or of fermentation was unknown.

But scientific attitudes were much more widespread and diffused than scientific knowledge. Attitudes of challenging traditional intellectual authority, deciding lines of development by systematic observation, measuring, testing, comparing, experimenting and adopting – indeed actively stimulating the development of – scientific devices such as the thermometer and hydrometer, which enables industrialists to reduce their empirical practices to rules wherever possible, were certainly being strengthened. The quest for more exact measurement and control, with research for the means to fulfil it, was certainly characteristic of these linkages – as with the introduction of thermometers into the pottery industry, thermometers and hydrometers in brewing and distilling – even where the objective was not to subvert empirical techniques, of which the chemistry remained unknown, but to standardize best practice within them. In this sense the developing Baconian tradition of the experimental sciences, the traditions of research based upon systematic experimentation (as in late eighteenth-century chemistry), had closer links with the process of innovation than did advances in cosmology, mechanics or physics. One might argue that the greatest gift of the scientific revolution to the industrial revolution was scientific *method* rather than scientific *knowledge*. And in such linkages science probably learned as much from technology as technology from science until the nineteenth century: scientists were much concerned with trying to answer questions suggested from industrial techniques.

In many ways the advance of scientific knowledge had virtually no direct effect upon technology. In mechanics, for example, very great theoretical sophistication was achieved in ballistics; but this left virtually untouched the processes of innovation in making metals or working metals, in gun-founding or the practice of gunnery. It needed a precision engineering industry using steel as the basic material and a precision chemical industry producing propellants before the technical sophistication of ballistics could be of much practical use. And this did not happen until after the Crimean War. A parallel case exists in the growth of the science of geology which, judged by results, had very little effect, or even connection with, the mining industry. The biblical controversies concerning the age of the earth inspired more geological advance than coal. The same is broadly true of stress theory and civil engineering practices in Britain, in the design of bridges and structures. Such formal knowledge was much more advanced in France. But, indirectly, when one looks at the general nature of the society and its intellectual parameters within which industrial advance was going on, rather than at the immediate context of innovation, there is more to be seen. Together, both science and technology give evidence of a society increasingly curious, questioning, on the move, on the make, having a go; increasingly seeking to experiment, wanting to improve. So, much of the significance impinges at a more diffused level, affecting motivations, values, general assumptions, the mode of approach to problem-solving and the intellectual milieu, rather than a direct transfer of knowledge. This, rather than direct contributions or 'spin-off', may have been the prime significance of the new popularizers of science and technology, the encyclopaedias, the amateur institutions and scientific societies, the new educational movements patronized by merchants and industrialists, the intriguing links between radical non-conformist scientific and business groups.

There are a few exceptions to the 'empirical' tradition, but, in my view, they are not representative. James Watt was a superb product of the Scottish university world, who had attended Joseph Black's lectures on latent heat at Glasgow University and undertook specific formal laboratory experiments to determine the 'elasticity' of steam. Some of the industrialists and innovators in the bleaching industry, where sulphuric acid became the basis of the process, were chemists, and trained as such, like the famous Dr Roebuck who had a hand in so many ventures. But his doctorate was actually in medicine. Rare metals, such as platinum (which was put to use in mirrors and jewellery), were developed by skilled chemists. Usually, in Britain, this meant a link with Scottish universities, or even continental ones, rather than Oxford and Cambridge. Individual examples of trained chemists pioneering innovations, or industrialists using trained chemists to advise them, do not form a representative selection.

The importance of mathematics was more generalized and the spread of mathematical teaching (which became the dominant teaching subject at Cambridge) was probably of more general significance than that of science. It figured prominently in more formal theoretical techniques for navigation (and duly appeared in the syllabuses of the endowed schools in the ports which were to be seminaries of seamen). Simple mathematics became prominent in surveying techniques and civil and mechanical engineering, as calculations became gradually more formal under the stimulus of persons such as Smeaton. 'Conversion tables' became very important in the eighteenth century and were very profitable publications. There were simple mathematical tables of all sorts: navigation tables from astronomical observations, surveyor's tables, tables for calculating the value of metal in ore, tables for land agents and stewards, for millwrights, customs and excise gaugers, coopers, watchmakers, opticians. All embodied the simple application of mathematics to practical empirical problems.

The skills that underlay innovations cannot be ascribed, either, to any great increase in mechanical ingenuity. The highest standards of precision and ingenuity known in the mechanical world belonged not to the industrial context directly at all, at least in strategic industries like iron and textiles. They flourished in scientific instrument-making, which belonged mainly to the world of the Royal Society, astronomy, botany and the Admiralty; in gun-making (with high precision standards in the workshops of Woolwich arsenal, for example); in watch and clockmaking and mechanical toys – the world of the luxury market. Very high standards of accuracy had been established in these areas of technology in the late seventeenth century and long before in the case of clockwork devices, such as cathedral clocks and performing automata, and they were as high in Paris or even St Petersburg as in London.

In Britain such skills were greatest, and most concentrated, in London, the site of the principal luxury and military markets, rather than in the provinces where most industry was located. One does see in the eighteenth century the development of important pools in these 'mechanical' skills in the provinces, notably in the Birmingham and Sheffield trades and in the Lancashire watch and clockmaking industry. Watchmaking, in particular, had become a large trade by the late eighteenth century, with as many as 120,000 watches being made annually in London, and almost as many in Lancashire. Watch and clockmaking exhibited an extreme division of labour (much more so than in Adam Smith's famous example of pin-making) and a range of specialist miniature precision tools, such as slide-rests, screw-cutting lathes and pre-set lathes, to ensure accuracy of reproduction in interchangeable parts. All could be fitted in a vice on a work-bench. Essentially the development of the mechanical engineering and machine-making industry of the late eighteenth century saw the translation of the precision standards existing in working brass – an easily worked metal, which could be worked cold, but which was not strong enough for large machines – and metals on a miniature scale into the large-scale working of iron, and then putting power to the machines. The maximum motive power for the early precision was clockwork and the steel spring; human muscle worked all the precision tools. Each unit produced was therefore expensive, destined still for the limited luxury market, with productivity per head low, even though production was rationalized by division of labour. The engineering industries of the late eighteenth century thus saw a meeting point between two older sorts of skills, which were both widespread before the industrial revolution: those of the blacksmith, carpenter, wheelwright and millwright being taken nearer to the standards of accuracy known by the clockmaker. Carpenters, metal workers and clockmakers formed the usual combination of skills demanded in advertisements for machine-makers in late eighteenth-century newspapers. It was the achievement of organizers of workshops like Matthew Boulton in Birmingham and Henry Maudsley later in London to take in millwrights and carpenters and clockmakers and turn them out as fitters. But this merging of skills followed the momentum for industrial innovation created by rising demand – the ingredients of these mechanical skills predating the changes in technique.

The development of steam power

Steam power invites a particular mention in the discussion of the springs of innovation and the place of science. It has often been seen as the greatest gift from science to industry in the eighteenth century, born and developed exactly in the world of an international competition among amateur scientists and their leisured patrons in the late seventeenth century; the precise linkage between university science and innovation being carried on by Watt from the 1760s – the classical example of scientific knowledge in alliance with commercial motivations.

Agreed. But there are complications. The advances in efficiency at the different 'stages of growth' of the innovation were strongly affected by other criteria, notably the problems of construction, of standards of precision in metal-working that alone made effective use possible on a commercial basis, of getting steamproof valves and joints (which required accurate plane surfaces), fitting a piston to a cylinder throughout its length and getting accurately cast parts (which would now be machined). Final tolerances in all these devices depended upon the handicraft skills of artisans, even where machine tools such as lathes existed. These skills, it can be argued, more than anything else set the limits of efficiency in operational terms, and these efficiencies came from the empirical world of John Wilkinson and Matthew Boulton, with rising standards of its own, a world increasingly working to rules, but still innocent of formal scientific thought.

The other complication is that Boulton and Watt were far from alone in producing steam engines during the last twenty years of the eighteenth century, even though their patents were slowing up the pace of diffusion of steam-power technology. By 1800, when the patents were thrown open, their engines were among the most old-fashioned – even if still by far the most reliable thanks to the unrivalled standards possessed by Boulton's workshops. Watt's patents and his own conservatism, born of success, were alike responsible. Boulton had pushed him into taking out the patents for rotative motion in 1781, on the arguments that there was only one Cornwall and that manufacturers elsewhere were 'steam-mill mad'. Watt was reluctant to experiment with road or rail traction, or engines for boats. He set his face against experimenting with 'high-pressure steam'. By the 1790s, a score of engineers in Cornwall, Lancashire, Birmingham, Northumberland, Scotland and London – almost every major manufacturing and mining region – were at work in all these extensions of steam technology. Trevithick, Hornblower and Stevenson were only the most famous of these. None of these innovations, with the gradual 'continuum' style improvement in the Watt-engine itself, was like Watt's, and none of these engineers was in the same scientifically literate, scientifically trained tradition as Watt; they belonged, for the most part, to the empirical world of the obscure colliery engineer, the captain of a Cornish mine, the brilliant mechanic such as William Murdoch, who were not seeking to create their devices of improvements in the light of awareness of scientific fundamentals. Yet the cumulative effect of 'continuum innovation', effected on the job, bit by bit, were profound.

It happened that crude measurements were taken of the efficiency of the steam-pumping engines from the year of their introduction in 1712.[5] Not accidentally, this took place in Cornwall, one of the only mining areas far from a coalfield, where the cost of fuel was high and the fuel-efficiency of the engine therefore worth measuring by way of its 'duty count'. The first Newcomen engine had a 'duty' – that was the number of pounds of water raised one foot by the consumption of one bushel of coal – of about 4.5 million. This had been raised to about 12.5 million by the time of Smeaton's improvements in the 1770s. The initial Watt separate condenser engine made a jump to about 22 million. By 1792 continuing improvement, particularly through making the engine 'double-acting' (where steam pressure was applied to both sides of the piston), had brought the duty counts to over 30 million. Average duty rates in low-pressure Cornish beam engines (i.e. Watt-style engines) then quadrupled to almost 100 million under 'continuum-type' improvements by the 1840s. This is not to say that

---

[5] D.B. Barton, *The Cornish Beam Engine* (Truro, 1965), pp. 28–59.

such gradual improvements were more strategic than the bigger, identifiable breakthroughs. Measured *ex post facto*, their net contribution may have been greater, but they may have been largely tributary to the major innovations, and would not necessarily have occurred in the absence of major innovations. A 'strategic' blockage on a small front, at the advancing edge of technology, could hold up innovation in a very much wider span behind it.

Select bibliography

General studies

T.K. Derry and T.I. Williams, *A Short History of Technology* (Oxford, 1960), Part II; D.S. Landes, *The Unbound Prometheus* (Cambridge, 1969); P. Mathias, *The First Industrial Nation* (London, 1983), ch. 5; N. Rosenberg, *Perspectives on Technology* (Cambridge, 1977) and *Inside the Black Box* (Cambridge, 1982); C. Singer (ed.), *A History of Technology* (Oxford, 1958); A.P. Usher, *A History of Mechanical Invention* (Cambridge, Mass., 1954).

Resources

T.C. Barker, 'Lancashire coal, Cheshire salt and the rise of Liverpool', *Transactions of the Historic Society of Lancashire and Cheshire*, LIII (1951); M.W. Flinn, *The History of the British Coal Industry Vol. II: 1700–1830 The Industrial Revolution* (Oxford, 1984); R.P. Multhauf, *Neptune's Gift: a History of Common Salt* (Baltimore, 1978); K. Warren, *Chemical Foundation: the Alkali Industry in Britain to 1926* (Oxford, 1980); E.A. Wrigley, 'The supply of raw materials in the industrial revolution', *Economic History Review* XV (1962), reprinted in *People, Cities and Wealth* (Oxford, 1987), ch. 4.

Theory and concepts

N. Rosenberg (ed.), *The Economics of Technical Change* (Harmondsworth, 1971); E. Mansfield, *The Economics of Technological Innovation* (New York, 1968); S.C. Gilfillan, *The Sociology of Invention* (Cambridge, Mass., 1970); J. Schmookler, *Invention and Economic Growth* (Cambridge, Mass., 1966).

Diffusion of innovations

K. Bruland, *British Technology and European Industrialisation: the Norwegian Textile Industry in the Mid-nineteenth Century* (Cambridge, 1989); J.R. Harris, 'The diffusion of English metallurgical methods to 18th century France', *French History*, II (1988); 'Attempts to transfer English steel techniques to France in the 18th century', in S. Marriner (ed.), *Business and Businessmen* (Liverpool, 1978); W.O. Henderson, *Britain and Industrial*

Europe 1750–1870 (Leicester, 1965); *The Industrial Revolution on the Continent* (London, 1967); D.J. Jeremy, *Transatlantic Industrial Revolution: the Diffusion of Textile Technologies between Britain and America 1790–1830* (Oxford, 1981); P. Mathias, 'Skills and the diffusion of innovations from Britain in the eighteenth century', in P. Mathias, *The Transformation of England* (London, 1979), ch. 2; E.H. Robinson, 'The early diffusion of steam power', *Journal of Economic History*, XXIV (1974); J. Tann and M. Brechin, 'The international diffusion of the Watt engine 1775–1820', *Economic History Review*, XXXI (1978).

Science and technology

Derek Hudson and K.W. Luckhurst, *The Royal Society of Arts 1754–1954* (London, 1954); P. Mathias (ed.), *Science and Society 1600–1900* (Cambridge, 1972); Jack Morrell and Arnold Thackray, *Gentlemen of Science* (Oxford, 1981); A.E. Musson and Eric Robinson, *Science and Technology in the Industrial Revolution* (Manchester, 1969); A.E. Musson (ed.), *Science, Technology and Economic Growth in the Eighteenth Century* (London, 1972); R.E. Schofield, *The Lunar Society of Birmingham* (Oxford, 1963); A. Thackray, 'Natural knowledge in cultural context, the Manchester mode', *American Historical Review*, LXXIV (1974).

Steam power

D.S.L. Cardwell, *From Watt to Clausius: the Rise of Thermodynamics in the Early Industrial Age* (London, 1971); R.L. Hills, *Power in the Industrial Revolution* (Manchester, 1970); J. Payen, *Capital et machine à vapeur au XVIIIe siècle* (Paris, 1969); E. Robinson and A.E. Musson, *James Watt and the Steam Revolution* (London, 1969); G.N. von Tunzelmann, *Steam Power and British Industrialisation* (Oxford, 1978).

More detailed passages on natural resources, technological change and innovation may be studied in monographs devoted to specific industries, which are too numerous to list in this selective bibliography.

3
Revisions and Revolutions: Technology and Productivity Change in Manufacture in Eighteenth-Century England

Maxine Berg

Interpretations of the industrial revolution

We used to look back to the industrial revolution as the great turning point in our history, as the origin and indeed cause of modern society. Eric Hobsbawm described it in his great classic, *Industry and Empire*, published in 1968, as 'the most fundamental transformation in the history of the world recorded in written documents. For a brief period it coincided with the history of a single country, Great Britain.'[1] His words convey images of new technology and industry, the steam engine and the cotton mill. This industrial revolution was a Prometheus. This is not unlike the perceptions of those in the 1830s and 1840s who saw themselves as living through an Age of Machinery. Thomas Carlyle spoke of 'the huge demon of Mechanisation ... changing his shape like a very Proteus ... and infallibility at every change of shape, oversetting whole multitudes of workmen.'[2]

Recently, however, historians have turned to a much more gradualist industrial revolution, seeing it as a phenomenon stretching back to the early days of the eighteenth century and continuing at least until

---

[1] Eric Hobsbawm, *Industry and Empire* (Harmondsworth, 1978), p. 13.
[2] Thomas Carlyle, cited in Raymond Williams, *The Long Revolution* (Harmondsworth, 1965), p. 88.

the mid-nineteenth century. This perspective has dominated historiography in varying degrees since the early 1970s.³ The latest quantitative work on the period takes this view much further; its effect has been to dethrone the industrial revolution altogether. Within the past five years a whole series of new indices has replaced the much used estimates provided by Deane and Cole in their *British Economic Growth*. Wrigley and Schofield's estimates on population increase have pushed the beginning of population change back to the very early eighteenth century, and have charted an upward movement of population over the whole period. Harley, McCloskey, Crafts and Williamson have produced new indices on growth rates and productivity. O'Brien has marshalled price indices of agricultural and manufactured commodities, and Lindert has constructed new indices on the social and occupational distribution of the labour force.⁴

This chapter will address the findings of the new picture of the industrial revolution which relies primarily on aggregative indices of economic growth.⁵ It assesses the findings and analysis of this new economic orthodoxy in comparison to those espoused earlier under the aegis of the estimates of Deane and Cole. It examines early criticisms of the new view. It then discusses two problems of interpretation which cast doubt, I think, on the conclusions of the

---

³ See David Cannadine, 'The past and present in the industrial revolution', *Past and Present*, 103 (1984), pp. 149–58.

⁴ See N.F.R. Crafts, *British Economic Growth during the Industrial Revolution* (Oxford, 1985); C.K. Harley, 'British industrialisation before 1841: evidence of slower growth during the industrial revolution', *Journal of Economic History*, 42 (1982), pp. 167–89; P.H. Lindert and J.G. Williamson, 'English workers' living standards during the industrial revolution: a new look', *Economic History Review*, 36 (1983), pp. 1–25; P.H. Lindert, 'Why was British growth so slow during the industrial revolution?', *Journal of Economic History*, 44 (1968), pp. 689–712; N.F.R. Crafts, 'The new economic history and the industrial revolution', in P. Mathias and J. Davis (eds), *The First Industrial Revolutions* (Oxford, 1989), pp. 25–43; D.N. McCloskey, 'The industrial revolution: a survey', in R.C. Floud and D.N. McCloskey (eds), *The Economic History of Britain since 1700*, vol. 1 (Cambridge, 1981), pp. 103–27; E.A. Wrigley, 'The growth of population in eighteenth century England: a conundrum resolved', *Past and Present*, 98 (1983); P.K. O'Brien, 'Agriculture and the home market for English industry, 1660–1820', *English Historical Review*, 100 (1985), pp. 773–800; J. Mokyr (ed.), *The Economics of the Industrial Revolution* (Totawa, NJ, 1985).

⁵ Another critique of the aggregative approach to the industrial revolution, written separately but contemporaneously with this one, and raising some similar issues in the introductory section, can be found in Pat Hudson, 'The regional perspective', in P. Hudson (ed.), *Regions and the Industrial Revolution* (Cambridge, 1989). I am grateful to Pat Hudson for allowing me to see her paper, and for our discussions, which have been very helpful in the revisions of this chapter.

quantitative historians: first their dependence on dualistic models of economic development, and second their limited perspective on sources of productivity change in the industrial sector. The chapter looks in depth at technological change and labour force participation, arguing that productivity gains have been hidden by excessively narrow definitions of technology and the labour force. Industry experienced gains in productivity due to product and process innovations neglected by a narrow focus on capital-output ratios. Industry also gained from the use of a child and female labour force that is entirely missed in the current labour supply data used by the quantitative historians.

Revisions of Deane and Cole

The conclusions drawn from recent estimates must be set against the perspectives which prevailed when Deane and Cole's estimates were accepted. Deane and Cole, in *British Economic Growth*, presented an integrated picture of the eighteenth-century origins of the industrial revolution; they marshalled data on trade, both foreign and domestic, on industrial output and capital formation, and on agriculture and population growth to demonstrate parallel and inter-related growth in all these sectors. They also sought an explanation for the growth they described in the special dynamic produced by this conjunctural growth. They were particularly concerned to present an industrial transformation which went far back in the eighteenth century, at least as far back as the 1740s. Deane and Cole did not in fact present their work as particularly novel: they saw it rather as a revival of an older historiographical tradition. What they did see in their results was a refutation of the then popular views of 'Nef, Ashton and Rostow that the decisive turning point in the economy had to be brought forward from the 1760s to the 1780s'. They pushed the acceleration in growth further back to the 1740s, and argued for an even earlier phase of growth in the very early eighteenth century, one which faltered by the 1720s, only to be properly resumed in the 1740s. Their assessment of industrial output indicated 'that the acceleration of output in the last quarter century may have been less of a break with past trends than the earlier upsurge'.[6] Nor were they unaware of the wider perspective of the impact of the cotton industry. For this reason, they looked at the

---

[6] See Phyllis Deane and W.A. Cole, *British Economic Growth* (Cambridge, 1962), p. 40 and chapter 2.

textile industry as a whole and emphasized the predominant role of the woollen industry until the second decade of the nineteenth century – in 1805 the net value of output of cotton was only 3½ per cent of the output of the UK as a whole; value added to national income by all textiles was estimated at 11 per cent of that of the UK.

Deane and Cole furthermore tied their economic indicators to an attempt to analyse the dynamics of growth. Their analysis was based on the favourable conjuncture of expanding home markets, population growth and agricultural development; and not on foreign trade, high capital investment or industrial innovation. Their analysis of more long-term growth was subsequently reinforced by E.L. Jones, Stanley Chapman and others, who also offered substantial microeconomic analysis of agricultural progress and capital formation. A long slow process of indigenous agricultural change, and rather low rates of capital formation, with the emphasis on circulating not fixed capital, have thus become an accepted part of the canon of the industrial revolution. This was further reinforced by studies of productivity change which took the advance and pervasive impact of steam power, factory organization and machinery from the centre of the industrial revolution, putting in their place intermediate technical change, and the increasing intensity and discipline of labour.[7]

The novelty of the substance of the new interpretations offered by the quantitative historians is cast into perspective by these long-standing and indeed traditional assessments. David Cannadine has recently demonstrated the widespread existence, at least from the early 1970s of a 'limits to growth' school.[8] This school, in turn, was not new, but reached back to the 1930s in the work of Clapham, Redford and Lipson, who wrote of a slow and localized industrial revolution. The optimistic vision of a high growth, capital-intensive industrial revolution sparked by short-term take-off appears, in fact, to have been confined to the relatively brief interlude of the 1950s and 1960s, when many economic historians were heavily influenced by contemporary theories of economic growth.

While the quantitative historians do not, therefore, represent the complete break with traditional interpretations that they imagine, they have certainly gone beyond these. For Williams, Mokyr and especially

---

[7] See, for example, C.K. Hyde, *Technological Change and British Iron Industry 1700–1870* (Princeton, NJ, 1977); G.N. von Tunzelmann, *Steam Power and British Industrialisation to 1860* (Oxford, 1978); and V.A.C. Gattrell, 'Labour, power and the size of firms in Lancashire cotton in the second quarter of the nineteenth century', *Economic History Review*, 2nd series, 30, pp. 95–139.

[8] See Cannadine, 'The past and the present in the industrial revolution'.

Crafts have gone beyond slow growth to question the very existence of industrialization before the mid-nineteenth century. Their new quantitative estimates have now replaced Deane and Cole's indices. The Deane and Cole analysis of the process of economic growth has also been replaced by recent developments in economic analysis. The findings of the revisionists have quickly formed the basis for a new orthodoxy on the economy of eighteenth- and early nineteenth-century England.

This new orthodoxy is best presented by summarizing the findings of N.F.R. Crafts, who has brought a number of the new indices together, and added his own new estimates of per capita and aggregate growth. Crafts presents a much more pessimistic picture of the eighteenth century than we have ever previously had. He has criticized other revisionists, including McCloskey and Williamson, for exaggerating productivity. He estimates productivity growth in manufacturing at only 0.2 per cent in 1760–80, 0.3 per cent in 1800–30 and 0.8 per cent in 1830–60, and argues that productivity growth in manufacturing was probably very slow until 1830. One small and atypical sector – cotton – probably accounted for half of all productivity change in manufacturing. In Crafts's words, 'not only was the triumph of ingenuity slow to come to fruition, but it does not seem appropriate to regard innovation as pervasive'. Productivity advance, he argues, was much more important in agriculture at least until 1760.[9] Floud has effectively summarized the implications of Crafts's indices:

> Britain underwent the pain of structural change but without the reward of rapid income growth ... technical change came slowly and patchily, with the spectacular changes in textiles disguising the backwardness of many other sectors. Thus, although an extraordinary proportion of workers entered industry, Britain did not get as much from them as it should have done, principally because of a neglect of education and a concentration on producing low wage factory fodder rather than high wage technicians.

The upshot was that 'Britain has simply muddled through 200 years of economic growth'.[10]

Crafts's indices suggest that before 1760 national production probably increased at the same rate as indicated by Deane and Cole, but

---

[9] Crafts, *British Economic Growth during the Industrial Revolution*; and 'The new economic history and the industrial revolution'.

[10] Roderick Floud, 'Slow to grow', review of N.F.R. Crafts, *Times Literary Supplement* (19 July 1985), p. 794.

after 1780 the estimates diverge sharply. Crafts found a substantially lower growth in output per head in the later part of the century. While Deane and Cole had envisaged a marked acceleration at that time. Even more dramatically, Crafts's figures showed a substantial decline in the rate of growth in product per head in the period 1760–1801 compared to that of 1700–60.[11]

Crafts's estimates of slow growth during the classic industrial revolution have been confirmed by other quantitative economic historians. Mokyr refers to a 'consensus among the proponents of the "new view" that in the first half century or so of the Industrial Revolution its economy wide effects were limited and economic growth was rather slow'.[12] Williamson speaks even more confidently:

> The quantitative dimensions of the classic British Industrial Revolution are understood far better now than a century ago. . . . Informed guesses on the rate of total factor productivity growth are now available, and as we have seen . . . even trends in workers' living standards have been nailed down securely. . . . Most of us now agree that British growth was slow up to about 1820, and much faster thereafter . . . we also agree that the rate of accumulation was slow throughout the first Industrial Revolution.[13]

The new orthodoxy is, however, itself already the subject of debate. Jackson has recently pointed out problems in the measurement of government expenditure, and he has argued that the effect of this on the measured rate of growth of total and per capita output is substantial.[14] The effect of different measures of government expenditure is to lower Crafts's estimate of national productivity growth before 1760, and to raise it for the period 1760–1800. Jackson argues that Crafts's finding that per capita growth was lower in 1760–1801 than in 1700–60 is due partly to the treatment of government expenditure, and partly to a computational error affecting 1780–1801.

Although Williamson and Mokyr agree with Crafts's scenario of

---

[11] This is also summarized in R.V. Jackson, 'Government expenditure and British economic growth in the eighteenth century: some problems of measurement', Discussion Paper, ANU (1988), p. 3.

[12] Joel Mokyr, 'Has the industrial revolution been crowded out? Some reflections on Crafts and Williamson', *Explorations in Economic History*, 24 (1987), p. 293.

[13] Jeffrey Williamson, *Did British Capitalism Breed Inequality?* (London, 1985); Williamson, 'Debating the British industrial revolution', *Explorations in Economic History*, 24 (1987), p. 269.

[14] R.V. Jackson, 'Government expenditure and British economic growth in the eighteenth century: some problems of measurement', *Economic History Review* (1990).

slow growth, they differ over its explanations and its timing. Crafts explained slow growth by low rates of total factor productivity growth, which in turn caused low rates of accumulation. Slow growth was caused by supply side considerations. Private capital formation was not crowded out by war debt, but as a pioneer industrializer Britain found it hard to achieve high rates of productivity growth on a wide front. Only a gradual acceleration in growth was available. Williamson, in contrast, emphasized the constraints of saving on accumulation, arguing that the increase in government debt during the eighteenth century onwards from 1776 to 1815 reduced investment and slowed down capital formation, thus reducing output and the growth of consumption. Williamson furthermore identified 1820 as a key turning point for both growth and increases in the standard of living; before this time the expansion of war debt was largely responsible for Britain's low rate of accumulation.[15]

Mokyr believes that though Williamson has probably overstated the effects of crowding out on accumulation, nevertheless he was right to argue that the wars exercised a major effect. They certainly caused supply shocks, seen through higher transportation and transaction costs. He argues that in spite of exaggerations, Williamson was probably closer to the truth than Crafts, who hardly mentions the effect of war.[16] Mokyr is also critical of the excessive significance attached by Crafts to estimates of agricultural total factor productivity, and of his downgrading of opportunities for improvements in industrial productivity.[17]

These differences among the revisionists lie both in aggregate indices and in their interpretation. Some of the revisionists have pointed out the contingency of their data, the restrictive assumptions, the limitations of their models, the large margins of error, and the alternative techniques of measurement which throw up different conclusions. Aggregate estimates carry particularly high margins of error. As Feinstein has pointed out, estimates of national capital and flows of capital, 'in a few cases ... especially before 1800 ... rely on fragments of evidence glued together with rough guesses and more or less arbitrary assumptions'.[18] Lindert concedes high margins of error in his estimates of occupational distribution. He reckons that for his finer

---

[15] Williamson, 'Debating the British industrial revolution', pp. 286–94.
[16] Mokyr, 'Has the industrial revolution been crowded out?', p. 302.
[17] Ibid., pp. 308, 312–15.
[18] Charles Feinstein, 'Capital accumulation and the industrial revolution', in Floud and McCloskey, *The Economic History of Britain since 1700*, p. 129.

occupational grouping (fewer than 40,000) the true numbers could be one-third to three times his estimates (at the 95 per cent level of subjective confidence). Estimates for shoemakers, carpenters etc. were 'little more than guesses'. And for categories with over 100,000 persons (agriculture, commerce, manufacturing etc.), the true value could be three-fifths to five-thirds the estimate.[19] Mokyr concludes: 'it seems we have run into strongly diminishing returns in analysing the same body of data over and over again . . . the highest return strategy now is to uncover new data.'[20]

Problems with the new orthodoxy arise not just from continuous analysis of the same body of data, but also in explaining sources of slow productivity change. Crafts's slow growth economic profile turns to a considerable extent on a reduction of the contribution of the industrial sector. Instead he sees agriculture as an engine of growth. 'Its performance permitted the migration of labour to the industrial sector, it kept food prices under control, and it accounted for most of the [slow] growth before 1830.'[21] I will therefore now turn to a consideration of the industrial sector and productivity change in it.

Crafts's findings of low productivity in the industrial sector are backed up by an assumption that technology and work organization were primitive in all but his 'glamour sectors' of cotton and iron. It is interesting that, keen though he is to dissociate the term industrial revolution from rapid transformation to factories and machinery, his model of technical change is exclusively contained within this framework. The recent theory of proto-industrialization, fraught though it is with problems, both theoretical and applied, has at least highlighted the role of change associated with handicraft production, particularly rural putting out. It identified an innovation in organization whose success lay in extracting a greater surplus out of an extensive and flexible labour force. Other broadly based research in economic history, and the sociological research focused recently on 'flexible specialization', has pointed out the gains of market and product development and technological innovation in the small scale sector.[22] Explanations of productivity change thus need to be sought beyond the aggregate data for contributions of capital, labour and total

---

[19] Lindert, *Journal of Economic History*, 1968, p. 701.
[20] Mokyr, 'Has the industrial revolution been crowded out?', p. 318.
[21] Ibid., p. 305.
[22] See Maxine Berg, *The Age of Manufactures* (London, 1985), chapters 8–12; and C. Sabel and J. Zeitlin, 'Historical alternatives to mass production', *Past and Present*, 108 (1985), pp. 133–76.

factor productivity. They must be analysed at source in the world of work and its social and cultural contexts. The limited assessment given the industrial sector in the new orthodoxy relies on a series of models themselves requiring reassessment.

Dualism and the industrial revolution

The basic model deployed by most of the revisionists is a dual economy model, that is to say, a division of the economy into traditional and modern parts. Crafts has created several variations on this divide. He argues that his productivity estimates are for industry rather than just manufacturing. They include mining, building and handicrafts. He elaborates on his earlier distinction between 'glamour manufacturing sectors and traditional industry', or between 'traditional' and 'revolutionized' sectors[23] to draw a distinction between 'production of goods for large dispersed national or international markets, and the production of goods and services for a local market'.[24] Once it was clarified how much manufacturing, mining and building was in handicrafts and non-tradeables, he thought it was certainly plausible for productivity growth in agriculture to be faster than in industry. He found the implied rate of technical progress in the unmodernized sectors to be approximately zero, and contrary to McCloskey's view, 'technical progress was not pervasive at a rapid rate through the Industrial Revolution'.[25] Crafts has argued that instead there was unbalanced productivity growth within industry; that is, that Britain had relatively rapid growth of productivity in traded goods, both goods generally exported and those imported, as compared with goods sold only in home markets. There were big productivity differences between the progressive parts of the economy, including agriculture, transport and a subset of industry, and the rest.[26]

Mokyr also proposes a dual approach, distinguishing the traditional from the modern parts of the economy. But his definitions of the modern and the traditional clearly differ from those of Crafts. He includes in his modern section all factories, transport, mining, quarrying,

---

[23] See Crafts, *British Economic Growth*, p. 17.
[24] Crafts, 'British economic growth 1700–1850: some difficulties of interpretation', Paper to the Workshop on Quantitative Economy History, University of Groningen (September 1985), pp. 3–4.
[25] Ibid., p. 255.
[26] Ibid., pp. 255–6.

metallurgy, paper and potteries. His traditional sector includes agriculture, domestic industry, food-processing, small scale metalworking shops, construction and building. He argues that slow growth of output per person hour in the traditional part diluted aggregate growth until the modern sector became large enough to dominate the movement of the economy.[27]

The dualistic division of the economy, however defined, is accepted by Crafts and Mokyr in a way closely tied to Wrigley's pessimistic view of the early industrial economy. As Williamson puts it, Crafts's estimates confirmed the classical economists' pessimism that between 1761 and 1831 the difference between the rate of growth of capital and that of output was trivial and failed to offset the impact of increasing land scarcity.

Wrigley accepts this classical pessimism, taking an entirely Malthusian perspective on the economy before the mid-nineteenth century, and he draws on the classical economists to underwrite his views. He believes that the classical economists were right to argue as they did; indeed, 'their reluctance to envisage the possibility of large gains in individual productivity find support in ... Crafts' estimates ...'. Their systems were dominated by 'negative feedback loops', and most of the economic changes taking place until the 1830s and 1840s are in his view best understood within the framework they posed of constraints of resource scarcity and population growth.[28] After this time, a new system emerged, largely because of the deployment of inanimate sources of energy and inorganic sources of raw materials.

> The natural technology of the day, though demonstrably capable of substantial development, especially under the spur of increased specialisation of function, was not compatible with the substantial and progressive increase in real incomes which constitutes and defines an Industrial Revolution ... the raw materials which formed the input into the production processes were almost all organic in nature, and thus restricted in quantity by the productivity of the soil.[29]

Power sources and raw materials were thus subject like agriculture to declining marginal returns to the land.

Wrigley confirmed these views in census data which showed that as late as 1831, in spite of comparatively high urban populations, most of

[27] Mokyr, 'Has the industrial revolution been crowded out?', p. 315.
[28] E.A. Wrigley, 'The classical economists and the industrial revolution', in *People, Cities and Wealth* (Oxford, 1987), pp. 22–34 and 36.
[29] Ibid., p. 9.

the labour force was working in occupations supplying goods for local markets. Those working in factories or proto-workshops made up only 10 per cent of the adult male labour force. While agricultural productivity increased substantially in the seventeenth and eighteenth centuries, releasing labour into other occupations, output per head in most of these did not improve a great deal. As late as 1831–41, at least two-thirds of the total increase in adult male non-agricultural employment was in occupations like building labourers, butchers, alehouse keepers, shoemakers, tailors, blacksmiths and bakers.[30]

The new orthodoxy thus divides the economy into the traditional and modern, and accepts the over-riding primitive state of most manufacture, defined as it is as part of the traditional part. Crafts and Mokyr have recently conceded that there may have been some limited technical progress in handicrafts. Wrigley accepts that framework knitting, coastal shipping, some forms of metal goods manufacture, brewing, glassmaking and papermaking 'may have constituted exceptions to the general rule'.[31] But all agree that these were no more than minor exceptions, and could never have provided the scale of cost reductions which were brought about in what they assign to the modern part.

Dual economics

The analysis of industrial productivity deployed by all the revisionists relies on dual economy models. We must ask just how appropriate these models are to eighteenth-century England, and to what extent we can accept their underlying assumption of technological stagnation in much of the industrial sector in the eighteenth and early nineteenth centuries.

The reliance of the revisionists on dual economy models is reminiscent of the early phases of development economics, which used explanations based on the idea of economics with a surplus of labour. These explanations were couched in terms of a rural/urban dichotomy, labour-intensive/capital-intensive divisions, a formal/informal distinction and other variations on the traditional/modern divide. The development economics which dominated planning for two decades after the Second World War looked to a policy of accelerated and large

---

[30] Ibid., pp. 11–15.
[31] Crafts, 'British economic growth: some difficulties', p. 255; Mokyr, 'Industrial revolution crowded out', p. 313; Wrigley, *People, Cities and Wealth*, p. 15.

scale industrialization. The economy was divided into dual sectors of traditional and modern parts, and it was argued that with the process of modernization, the traditional parts of the economy would be absorbed into the modern. The traditional/modern divide was overthrown in the 1970s in favour of an informal/formal sector division and a reassessment of the role of small scale activities. Instead of a pool of stagnant disguised unemployment, informal activity was credited with essential urban services. 'From being the Cinderella of underdevelopment, the informal sector could become a major source of future growth.'[32] New attention to the informal sector only underlined the divide. Formal and informal sectors were seen as 'two juxtaposed systems of production, one derived from capitalist forms of production, the other from the peasant system'; as two types of economy, 'a firm centred economy' and a 'bazaar type economy'; as 'two sectors, a high profit/high wage international oligopolistic sector, and a low profit/low wage competitive capitalist sector'.[33]

From the mid-1970s this optimistic perspective on the informal sector was subjected to new scrutiny, which stressed the constraints on its expansion and the need for assessing potential dynamics in terms of the structural position of each sector. Subsequently, the validity of the whole dualist model was challenged. There was recognition of extensive internal differentiation in the urban economy, and an alternative framework was developed based on a continuum of economic activities rather than a two sector divide. Workers were seen as employed in a number of different categories, as self-employed, casual and wage and non-wage family labour. Recognition also went to the diverse dependent linkages between the activities of the 'traditional', 'handicraft', 'local market' or informal sector and that of the 'modern', 'factory', 'international' or formal sector. There were linkages through subcontracting, outworkers, the use of retail agents between the sectors and many more connections. Research discovered the connections between homeworking and sub-contracting carried out in our own inner cities and the peripheral urban areas of South-east Asia and Latin America on the one hand, and large international firms employing high-tech large scale factories on the other. The current Italian firm Benetton combines the use of domestic sub-contractors as a part of the

---

[32] Caroline Moser, 'Informal sector of petty commodity production: dualism or dependence in urban development?', *World Development*, 16 (1978), p. 1052; also see John Toye, *Dilemmas in Development, Reflections on the Counter Revolution in Development Theory and Policy* (Oxford, 1987).

[33] Ibid., p. 1052.

informal sector and high-tech production processes in the formal sector to balance its control of production, the market and distribution. Dualistic models were eventually found to be incapable of handling the complexities of these relationships.[34]

The dualism which was once fundamental to development economics has been overthrown, but it is now raised to new heights in economic history. The divide between agriculture and industry has been challenged by the revisionists who, contrary to older theories, now identify agriculture with substantial productivity gains while associating much industrial employment with nil productivity growth. With their recognition of the literature on proto-industrialization, they have now also divided the rural and the industrial sectors into factory and handicraft or putting-out. But in spite of such concessions, they still see the major divide as that between high productivity factory and mechanized industry along with a high productivity agriculture on the one hand, and a widespread industrial and service backwater on the other. In spite of concessions to some technical innovation in the handicraft sector, the divide between the modern and the traditional remains firmly entrenched. It accords with the industrial dualism that Piore describes as one that associates craft production and the small firms as complementary to mass production, but necessarily subordinate to and derivative of it.[35]

The divide between the modern and traditional sectors, sectors dealing in tradeables or non-tradeables, handicraft and factory industry is a convenient one for economists, but one that distorts the characteristics of manufacturing in the eighteenth century, and hides major sources of productivity gain in manufacture. Rigid associations of productivity gain and technical progress with concepts of large scale production, factories, powered machinery and capital-deepening pervade the revisionist position. In practice it was and is very difficult to make clear-cut divisions between the traditional and the modern, the tradeable and the non-tradeable, as there were rarely separate organizational forms, technologies, locations or even firms to be ascribed to either. Eighteenth- and nineteenth-century cotton manufacturers typically combined steam-powered spinning in centralized factories with large scale employment of domestic hand-loom weavers using traditional techniques. The small metalworking shops of Birmingham, Sheffield and Lancashire were classified by the revisionists with the traditional sector, handicrafts and non-tradeables although they

---

[34] Ibid., p. 1056.
[35] Sabel and Zeitlin, 'Historical alternatives to mass production', p. 138.

typically developed their high technology in the luxury goods trade of the home market, and also tried to break into and extend foreign markets. Artisans in the sector frequently combined occupations or changed these over their life cycle in such a way that they too could be classified in both the traditional and the modern sector. Firms primarily concerned with metalworking (classified by the revisionists as traditional), also diversified into metal-processing ventures (classified as modern), as a way of generating steady raw material supplies. The aggregate divisions between sectors which form the foundation of conclusions by the revisionists on the industrial sector are certainly questionable. Even more questionable is their understanding of the handicraft sector.

Handicraft and productivity

The revisionists associate the progressive with the machine and the factory: Wrigley places central emphasis on the need for inorganic sources of raw materials and power; Williamson on capital deepening; Crafts identifies the progressive sectors with tradeable goods; Mokyr finds the key transformation in modern mechanized industry which generated the dazzling cost reductions of the industrial revolution.[36] Wrigley's emphasis on inorganic materials and power would hold little sway in a comparative context: American and Swedish industrialization relied to a predominant extent on wood-using technologies and water power. Crafts, Mokyr and Williamson base their ideas of technological change on artefacts rather than processes. A broader concept of technological change would include not only machinery, but also tools, skills and dexterity, and the knacks and work practices of manufacture. And a broader definition of innovation must include product innovation, market creativity and organization change. The non-factory, non-tradeable and supposedly stagnant section of the economy experienced extensive technical change not recognized by the revisionists. Early textile innovations – carding and scribbling machinery, the Dutch loom, the knitting frame, the flying shuttle and the jenny, silk-throwing machinery and finishing techniques, especially in bleaching and calico printing – were all developed within rural manufacture and artisan industry, and few of these were initially

---

[36] Wrigley, *People, Cities and Wealth*, pp. 3–4; Williamson, 'Debating', p. 273; Crafts, 'British economic growth; difficulties', p. 254; Mokyr, 'Is the industrial revolution being crowded out?', pp. 314–15.

developed for the high profile cotton industry. The metalworking trades were proverbial for skill-intensive hand processes and hand tools. The stamp, press, drawbench and lathe were developed to innumerable specificiations and uses, and new malleable alloys, gilting processes, plating and japanning were at least as important. And other industries experienced some form of transformation in materials or division of labour, if not in the artefects of technological change.

The impact of these new techniques on productivity has never been adequately investigated. There have been some estimates based on contemporary opinion that the water-driven scythe hammer raised output per unit of labour to five times that of the hand forge; the treadle operated spinning wheel with a flyer increased productivity by one-third over that of the hand spinner. The flying shuttle doubled labour productivity, the Dutch loom increased labour productivity four-fold and the knitting frame ten-fold. Looked at in their own context, these are great gains, but they have generally been discounted against later developments in cotton-spinning techniques which soon made hundred-fold leaps over the old spinning wheels.[37] There are no productivity estimates for the range of hand tools and early machine tools in the metal industries, but their significance to overall productivity growth is a recurrent theme of economic history.[38]

A range of traditional industries certainly underwent reorganization due to changes in materials and processes. New industrial uses for coal affected brewing, brick-making, malting, sugar and soap-boiling as well as metallurgy and metalworking. Salt-refining based on rock salt solutions yielded ten times as much salt as natural brine solutions. The division of labour and production time of luxury industries such as hat-making and jewellery production were transformed by changes in materials, such as the replacement of beaver fur by hare skin or the introduction of silver plating and gilting.

In traditional textile industries changes in the product, such as the move from heavy serges to mixed stuffs, where wool was mixed with silk or cotton, considerably reduced the finishing time, for many of these needed no fulling, and they were dyed in the wool or printed rather than vat-dyed. The success of the calico printing industry later

---

[37] See P. Kriedte, H. Medick and J. Schlumbohm, *Industrialisation before Industrialisation. Rural Industry in the Genesis of Capitalism* (Cambridge, 1981), pp. 112–13; W. Endrei, *L'evolution des techniques du filage et du tissage du Moyen Age à la revolution industrielle*, Industrie et artisanat, vol. 4 (Paris, 1968).

[38] See Nathan Rosenberg, *Inside the Black Box: Technology and Economics* (Cambridge, 1982); Roderick Floud, *The British Machine Tool Industry*; David Landes, *The Unbound Prometheus* (Cambridge, 1969).

in the eighteenth century hinged not on new machinery or materials, but on new cheaper labour prepared to carry out labour-intensive processes on a new scale and under new organization and discipline.

The productivity benefits of these changes in processes and products are notoriously difficult to measure. For product innovation falls outside the conventional measures of productivity change, doomed as it is to the problem of assigning different weights to different products over time and for the effects of differential price charges. And changes in skills, organization and discipline may affect the quality of labour inputs without affecting the quantity of labour.[39] As contemporary literary evidence and current historical assessments make clear, for the eighteenth century at least, the expansion of consumption and the product innovation associated with that were the really essential elements of economic growth.[40]

Josiah Tucker's observation of 1757 was no figment of the imagination. And its sentiments, I have argued elsewhere, were widely echoed in the economic commentary of the eighteenth century:

> Few countries are equal, perhaps none excel, the English in the number of contrivances of their Machines to abridge labour. Indeed the Dutch are superior to them in the use and application of Wind Mills for sawing Timber, expressing Oil, making Paper and the like. But with regard to Mines and Metals of all sorts, the English are uncommonly dexterous in their contrivance of the mechanic Powers. Yet all these, curious as they may seem, are little more than preparations or Introductions for further Operations. Therefore, when we still consider that at Birmingham, Wolverhampton, Sheffield and other manufacturing Places, almost every Master Manufacturer hath a new Invention of its own, and is daily improving on those of others; we may aver with some confidence that those parts of England in which these things are seen exhibit a specimen of practical mechanics scarce to be paralleled in any part of the world.[41]

The revisionists argue that most industrial labour was, however, to be found in those occupations which really did experience little change. But these occupations, in the food and drink trades, shoe-

---

[39] See N. Rosenberg, *The Economics of Technological Change* (Harmondsworth, 1971); M. Salter, *Productivity and Technical Change* (Cambridge, 1960); W. Lazonick and T. Brush, 'The Horndal effect in early US manufacturing', *Explorations in Economic History*, 22 (January 1985).

[40] See John Brewer, Neil McKendrick, and J.H. Plumb, *The Birth of a Consumer Society* (London, 1982).

[41] Cited in Charles Wilson, *England's Apprenticeship 1603–1763*, 2nd edn (London, 1984), p. 311.

Revisions and Revolutions

making, tailoring, blacksmithing and trades catering to luxury consumption, were also a part of the unique urban expansion of early modern and eighteenth-century England.[42] They supplied the essential services on which town life was dependent, and provided for the remarkable flowering of a consumer culture in the eighteenth century, to which historians are now turning their attention.[43]

It is, furthermore, the case that early industrial capital formation and enterprise typically combined activity in the food and drink or agricultural processing trades with more obviously industrial activities. The separation of 'traditional' from 'modern' activities is an artefact of the modern economist, not a realistic analysis of the complexities of the eighteenth-century economy. This was true in textile manufacture[44] and in the metal manufactures in Birmingham and Sheffield where innkeepers and victuallers were frequently among the mortgagees and joint owners of metalworking enterprises. Such manufacturers also maintained joint occupations in the metal and food and drink trades. In the South Lancashire tool trades, Peter Stubs was not untypical when he first appeared in 1788 as a tenant of the White Bear Inn in Warrington. Here he combined 'the activity of innkeeper, maltster and brewer with that of filemaker'. And there were good technological reasons for this combination. One of the processes in filemaking involved covering the file with a paste to preserve it from damage: 'this paste consisted of malt dust and "barm bottoms" or the dregs of beer barrels. The carbon from these ingredients was made to enter the teeth of the file so giving it greater durability and strength.'[45]

Handicraft industry and the labour force

Another striking feature of the new orthodoxy is its restricted definition of the labour force, and this in turn is closely related to the

---

[42] See Wrigley's essays on urban growth in *People, Cities and Wealth*, pp. 133–57.

[43] J. Brewer, N. McKendrick and J.H. Plumb, *The Birth of a Consumer Society: the Commercialization of Eighteenth Century England* (London, 1982); T. Breen, 'Baubles of Britain: the American and British consumer revolutions of the eighteenth century', *Past and Present*, 110 (1988), pp. 73–105.

[44] See K.H. Burley, 'An Essex clothier of the eighteenth century', *Economic History Review*, xi (1958); Stanley Chapman, 'Industrial capital before the industrial revolution 1730–1750', in N. Harte and K. Ponting, *Textile History and Economic History* (Manchester, 1973).

[45] See. T.S. Ashton, *An Eighteenth Century Industrialist. Peter Stubs of Warrington 1756–1806* (Manchester, 1939), pp. 4–5; cf. Peter Mathias, 'Financing the industrial revolution', in Mathias and Davis, *The First Industrial Revolutions*, pp. 78–82.

analysis of productivity change. The literature on proto-industrialization highlighted the contributions to productivity increase made by the intensification and division of labour. It also drew attention to the age and gender differentiation of the labour force.[46] Substantial productivity increase was achieved in ways we rarely consider now by the special contributions of women's and children's labour.

Yet Wrigley assesses productivity growth only through the 10 per cent of adult male labour who in 1831 worked in industries serving distant markets. Williamson's documentation of inequality and Lindert and Williamson's survey of the standard of living consider only adult males. What place did female and child labour play in industrial employment? It is difficult to assign a quantitative estimate of this, for this is the part of the labour force which was excluded from official statistics. Lindert's estimates for industrial occupations rely only on adult male burial records. But the implications of the inclusion of child and female labour are significant. They dramatically effect the analysis of inequality. Williamson argued that 'demographic forces from below' did not account for the widening of inequality in the industrial revolution. But if incomes are viewed within the context of the household, then it must be argued that population growth put pressure on the wages of the male breadwinner, and also directly affected the supply of female and child labour and their wages. As Saito has argued, 'it is likely that sex differentials in wages also widened when population rose faster. Wages for unskilled males were perhaps stagnant, but wages offered to females and children may have been actually falling.'[47] Thus women's and children's labour appeared in the eighteenth century to be a lucrative source of profit not to be bypassed by manufacturers ready to launch new labour-intensive industries during the age of mechanization.

Although quantitative evidence on the amount of child and female industrial labour is sparse, there is enough to indicate the inadequacy of conclusions based only on adult male labour forces. Wrigley and Schofield estimate that children aged five to fourteen accounted for between 18 and 25 per cent of the total population. This compares to a

---

[46] See Hans-Medick, 'The proto-industrial family economy', *Social History* (October 1976); Maxine Berg, *The Age of Manufactures* (London, 1985), chapter 6; David Levine, 'Industrialisation and the proletarian family in England', *Past and Present*, 107 (1985).

[47] Osamu Saito, 'The other faces of the industrial revolution', Review Essay, Institute of Economic Research, Hitotsubashi University, 1988; also see his 'Labour supply behaviour of the poor in the English industrial revolution', *Journal of European Economic History*, X (1981), pp. 633–52.

proportion of 6 per cent in 1951. The employment of such large numbers of children would clearly be a major problem for any economy. Levine has argued that children could and did begin to pay their own way from an early age, thereby cushioning the impact of this high dependency ratio.[48]

Textiles were the most important manufacturing industry, accounting for 45 per cent of the increase in national product in 1770 and 42 per cent in 1801. Wool made up 30.6 per cent in 1770 and 18.7 per cent in 1801. In the woollen industry, women's and children's labour accounted for 75 per cent of the workforce; children's labour exceeded that of women and of men. Women and children also predominated in the cotton industry: children under the age of thirteen made up 20 per cent of the cotton factory workforce in 1816; those under eighteen 51.2 per cent.[49] The silk, lace-making and knitting industries were also predominantly female. There were even high proportions of women and children in metal manufactures, such as the Birmingham trades. Goldin and Sokoloff's data for America are more complete and show that women and children together grew from 10 per cent of the manufacturing labour force of the north-east early in the nineteenth century to 40 per cent in 1832.[50] Therefore, the fact that only 10 per cent of adult male labour was to be found in the modernized progressive sectors, as Wrigley points out, does not tell us a great deal. For the preferred labour force for precisely these sectors was overwhelmingly young and female.

Was this labour force a drag on productivity gain in those few industries the revisionists credit with any real growth? Or did this labour force have other attributes making it attractive to manufacture,

---

[48] See David Levine, 'Industrialisation and the proletarian family in England', *Past and Present*, 107 (1985). The argument is also elaborated in Levine, *Reproducing Families* (Cambridge, 1987).

[49] See Adrian Randall, 'The West Country woollen industry during the industrial revolution', unpublished PhD Thesis, University of Birmingham (1979), Vol. II, p. 249; Clark Nardinelli, 'Child labour and the factory acts', *Journal of Economic History*, 40 (1980), pp. 739–55. The gender and age differentiation of the eighteenth-century industry is discussed in more depth in Berg, 'Women's work, mechanisation and the early phases of industrialisation in England', in Patrick Joyce (ed.), *The Historical Meanings of Work* (Cambridge, 1987), pp. 69–76; and Berg, 'Child labour and the industrial revolution', Workship Paper to the University of Essex (1986). Also see Claudia Goldin and K. Sokoloff, 'Women, children and industrialisation in the early republic: evidence from the manufacturing censuses', *Journal of Economic History*, 42 (1982), pp. 421–774; and Claudia Goldin, 'The economic status of women in the early republic: some quantitative evidence', *Journal of Interdisciplinary History*, 16 (1986).

[50] See Goldin and Sokoloff, p. 743.

apart from its acceptance of lower wage levels? In the terms used by economists, were there supply side considerations apart from the labour demand effect of wage levels?

The significance of women and children to the manufacturing labour supply must affect estimates of occupational distribution between traditional and modern sectors. It must also lead us to enquire into the skills and attributes of this labour force. It is not sufficient to see this as a stagnant pool of unskilled labour. For manufacturers were attracted by this labour force not just for low wages, but also for labour supply characteristics which contributed to productivity gains. There is a recent analogy to this in research on the employment of women, especially young single school-leavers in the new manufacturing industries of the Third World. Women are selected rather than men for their docility and their 'nimble fingers'; the result for industrialists is low labour costs and high labour productivity.[51] In many eighteenth-century industries women and children were specifically sought out for their dexterity and amenability to discipline. They were, in addition, regarded as particularly suitable to a division of labour associated with eighteenth-century technologies, one based on adult labour with child assistants. Indeed there are several instances of early textile machinery being designed and built to suit the child worker. The spinning jenny was a celebrated case: the original country jenny had a horizontal wheel and required a posture most comfortable for children aged nine to twelve.[52] For a time in the very early phases of mechanization and factory organization in the woollen and silk industries, as well as cotton, it was generally believed that child labour was integral to textile machine design.[53]

This association between child labour and machinery may in England have been confined to a fairly brief period of factory

---

[51] See Ruth Pearson and Diane Elson, 'The subordination of women and the internationalisation of factory production', in K. Young, C. Wolkowitz and R. McCullagh (eds), *Of Marriage and the Market* (London, 1981), pp. 144–67; Ruth Pearson, 'Female workers in the First and Third Worlds: the greening of women's labour', in R.E. Pahl (ed.), *On Work* (Oxford, 1988), pp. 449–69; Kristine Bruland, 'The transformation of work in European industrialization', in Mathias and Davis (eds), *The First Industrial Revolutions*, pp. 154–69.

[52] These issues are explored in greater depth in my 'Child labour and the industrial revolution', Workshop on Child Labour and Apprenticeship, University of Essex, 1986; and in my 'Women's work and the early phases of mechanisation in England', in Joyce (ed.), *The Historical Meanings of Work*.

[53] See Select Committee on Children's Employment, vol. II, *Parliamentary Papers* (1816–17), pp. 279, 343; and Select Committee on Children in Factories, *Parliamentary Papers* (1831), p. 254.

development. In the USA, it appears to have been a much more straightforward development dating from 1812 and associated with new large-scale technologies or divisions of labour to dispense with skilled adult male labour.[54] In England, it seems that the early predominance of women and children in textile factories was a development arising from changes in technology and the division of labour taking place earlier in the eighteenth century under domestic and workshop production. The typical work group consisted of an adult and child assistant; the assistant saved time and helped to increase throughput. The system was in some cases dramatically expanded to workshops organized under hierarchical divisions of labour. Processes were, furthermore, broken down into a series of dexterous operations, which were performed particularly well by teenage girls who contributed learned manual dexterity and high labour intensity. The manufacturers who developed these new production processes and techniques in many eighteenth-century industries were particularly successful in perceiving and capturing the benefits of this labour force.

Access to cheap supplies of labour, especially that of women and children, was nothing new – it was integral to the spread of manufacture in the early modern period. But this was labour that was also endowed with special attributes particularly suited to eighteenth-century technologies and work organization. Young workers, as assistants when small and later as independent youth workers, and a large workforce of women workers provided a great boost to labour productivity. But their contributions have been bypassed in assessments of productivity which rely wholly on capital-labour ratios. The estimate for labour must be disaggregated into gender and age differences, skill and labour intensity. This is particularly crucial for the years usually assigned to the industrial revolution. It is likely there was a special place for this women's and children's labour in the phrases of industrialization that occurred in the eighteenth and very early nineteenth centuries. Some aspects of this phenomenon have only recently found a parallel in the decentralization of production processes in manufacturing in both the Third World and advanced industrial countries.

Conclusion

This chapter has examined aspects of the recent quantitative assess-

---

[54] Goldin and Sokoloff, p. 747.

ments of the industrial revolution. It expresses considerable doubts over the treatment of the industrial sector in these assessments. It has found major problems in the use made by the revisionists of dualistic models of the economy. It demonstrates an inappropriateness to the eighteenth-century economy of narrow and restrictive definitions of technological innovation. And finally, it examines the implications of entirely misconceiving the eighteenth-century labour force through relying on data for adult males only.

The kinds of technological and organizational innovation I have discussed in this chapter – those associated with product and materials innovation as well as the division and intensification of labour – are hidden behind capital-output ratios. They were also notably the most widespread and significant types of innovation that occurred in most sections of manufacturing in the classic years of the industrial revolution, that is between 1760 and 1820. The types of definition and measures adopted by the revisionists do not capture the effects of these forms of innovation, which were historically specific to the period. Neither do definitions of the labour force take sufficient account of the particular age and gender mix, which also seems to have been historically specific to these years. The emphasis of Crafts and others on continuity has masked the historical disjunctures which were probably distinctive for only a fairly brief period before more conventional and more easily quantified signs of industrialization appeared.

4

The Constraints of a Proto-industrial Society on the Development of Heavy Industry: the Case of Coal-mining in the South-east of France, 1773–1791

Gwynne Lewis

The debate surrounding the much-discussed – and much-abused – 'model' of proto-industrialization has hardly touched upon the problem of heavy industry. As Professor Clarkson writes in *Proto-industrialisation: the First Phase of Industrialisation?*, 'Practically all the examples are drawn from the woollen, linen and cotton industries'. He goes on to state that the reason for the neglect of industries like coal-mining is that, 'such forms of manufacturing do not fit into the dynamic aspects of the model'.[1] The same doubts were raised at the conference devoted to proto-industrialization that was held at Lille in 1981.[2] To what extent are these doubts justified? If one accepts Professor Clarkson's very broad description of proto-industrialization as pointing to 'a common feature of the economy of pre-industrial Europe: the existence of industries in the countryside',[3] then one

---

[1] L. Clarkson, *Proto-industrialisation: the First Phase of Industrialisation?* (London, 1985), p. 54.
[2] See the *Revue du Nord*, 63 (1981), pp. 5–251.
[3] Clarkson, *Proto-industrialisation*, p. 10.

should certainly not exclude mining, at least in its traditional form. However, the more precise features which do relate to Professor Clarkson's model of proto-industrialization tend to be those which explain the *constraints* placed upon 'revolutionary' industrial growth. This chapter will argue that modern industrial capitalism in the south-east of France during the last decades of the eighteenth century was impeded, rather than advanced, by the nature of a proto-industrial socio-economy, one which was embedded in a changing, but still powerful, seigneurial system. Thus, I shall be emphasizing the fact that, far from there being a linear relationship between traditional 'proto-industrial' and modern 'industrial' socio-economies, the former may well have exerted a negative force upon the emergence of modern industrial capitalism.[4] To support this argument I will examine: (a) the attempt, during the reign of Louis XVI, to introduce a modern capitalist form of production into the coal-mining industry of south-east of France; and (b) the reasons why this attempt failed. I shall close with some general remarks about the role of the French Revolution in the future development of coal-mining in France.

Geographically, I shall be concentrating on the Alès coal-basin, which is to be found in the mountains of the lower Cévennes valleys in the south-east of France. The town of Alès – the capital of the Cévennes – is the gateway to the mining valleys, whose pits have only recently stopped working. Alès, with a population of around 10,000 in 1789, forms the apex of a triangle completed by Montpellier, seventy kilometres to the south-west, and Nîmes, forty-four kilometres to the south-east. Beaucaire, site of one of the oldest and most important trading fairs in Europe, lies sixty-seven kilometres away on the left bank of the river Rhône. From there ships sailed to Marseille and the ports of the Mediterranean. The canal de Languedoc linked western Languedoc to the Rhône, but the stretch from Lunel to Beaucaire had not been completely finished before the Revolution. Thus, although Alès was relatively close to the Rhône and the Mediterranean, transport was never easy and always very costly. The cost was increased by the levying of heavy tolls on rivers like the Cèze, linking Alès with the Rhône, as well as on the river Rhône itself. This is the first – and not an insignificant – factor in the obstacles confronting entrepreneurs in *ancien régime* society, one in which the profits went primarily to the seigneurs who leased out the rights to collect such tolls.

[4] M. Berg, *The Age of Manufactures, 1700–1820* (London, 1985), p. 83, stresses the importance of 'plebeian culture' and its different 'code of rationality'.

Since the visit of Jean Calvin to Nîmes in the sixteenth century, the Cévennes region had been an arena of bloody religious strife between Calvinists and Catholics. The interminable wars of religion had been followed by the revocation of the Edict of Nantes at the end of the seventeenth century and then by the ferocious *Guerre des Camisards* during the first decade of the following century. Throughout the reign of Louis XIV, and the even longer reign of his successor, Louis XV (1715–74), Protestants, who represented strong numerical minorities in towns like Alès and Nîmes, and strong majorities in many of the smaller *bourgs* of the Cévennes, were persecuted and denied full civic rights, although the situation did improve later in the century, leading to an Edict of Toleration passed by the government of Louis XVI on the eve of the Revolution.[5]

Religious antagonisms played an unquantifiable part in the complex drama of socio-economic change, just as they do today in Northern Ireland, Iran, or Armenia, a fact frequently ignored by the more statistically inclined economic historian. Partly as a result of their – theoretical – exclusion from public office, Protestants had long dominated the financial and manufacturing life of the lower Cévennes, as well as the important silk manufacturing centres like Nîmes to the south. By the mid-eighteenth century, lower Languedoc had become one of the major manufacturing regions in France, indeed in Europe. The involvement of the bigger Protestant financial houses, however, was generally confined to financial, commercial and manufacturing affairs, some of the richest trading on a world scale.[6] Protestants were not to be prominent in the early stages of heavy industrialization, at least in the sphere of coal-mining. Investment finance had to be 'imported' from outside the province. Well into the nineteenth century, investment in land, in the production of wine and raw silk particularly, was to take precedence over risky industrial ventures like coal-mining or iron-smelting. It is worth stressing here that an obsession with the ownership of land, as far as the inhabitants of the Midi were concerned, was evident long before the Revolution, and was not unassociated with the crisis in the manufacturing sector after 1778. I shall return to this point.

The region covering the mines of the Alès coal basin was divided between two *seigneuries*. The *vicomté de Portes* was owned until the

[5] For a continuity of religious antagonisms in the Cévennes see G. Lewis, 'A Cévenol community in crisis: the mystery of "L'Homme à Moustache"', *Past and Present*, 109 (1985).

[6] M. Sonenscher, 'Royalists and patriots: Nîmes and its Sénéchaussée in the late eighteenth century', PhD thesis, Warwick University, 1978.

1780s by the powerful Conty family, but sold in 1783 to Monsieur, the king's brother. The Contys had always resisted the modernization and expansion of the mines under their control. The other *seigneurie* was the *comté d'Alais* (the spelling of the town was changed to Alès in the 1920s). This prestigious fief was one of the oldest in France, dating back to the noble, Raymond Pelet, who went on the First Crusade. Covering some fifty parishes in the lower Cévennes region, and including the right to levy feudal dues in and around the town of Alès, the *comte d'Alais* entitled the owner to a seat on the *Etats de Languedoc*. Here we have the first link between the social and economic and the political power of the nobility. As we shall see, the *Etats de Languedoc*, controlled by prelates and nobles, was to play an important role in defeating the schemes for the modernization of the coal-mines in the Alès basin before their abolition in 1789. Both *seigneuries* were empowered with rights of seigneurial justice, providing lucrative financial returns for local lawyers and *juges des seigneurs*.[7]

There is one final point by way of introduction – the almost total dependence of the manufacturing region of lower Languedoc upon imports of food, as well as other basic commodities like salt, from upper Languedoc to the north and the plain and ports of the Mediterranean coast to the south.[8] This complicated and, for the big wholesale merchants in textiles and grain (*négociants*), very lucrative *ancien régime* commercial and manufacturing interdependence, one which fits, as we shall see, quite neatly into Professor Clarkson's proto-industrialization model, was to be threatened by the emergence of heavy industrialization after 1770.

It is important to note from the outset that every informed visitor who visited the coal-mines of the Alès coal-field in the late eighteenth century was struck by their potential. This was particularly true of what was to prove the richest mine of all, the Grand'Combe, which was situated on the *seigneurie de Trouilhas* in the *comté d'Alais*. The local representative of the international mining and coking company,

---

[7] The private papers of the maréchal de Castries, *Archives nationales*, 306 AP, contain a wealth of information relating to the stewardship of his estates in the lower Cévennes (see, in particular, 306 AP 496) as well as to his struggle with Tubeuf.

[8] R. Laurent and G. Gavignaud, *La Révolution française dans le Languedoc Méditerranéen* (Toulouse, 1987), p. 23: 'Le Bas-Languedoc ne vit que grâce à un vaste réseau de ravitaillement, mis en place par le négoce, qui puise les grains dans le Haut-Languedoc, par les canaux des Deux-Mers et des Etangs; en Bourgogne, par la voie rhodanienne; à l'étranger, par les ports de Marseille, Aigues-Mortes, Agde et surtout Sète.'

Richard, Carrouge et Cie, informed his main office in Paris in 1779: 'One might say that the coal owned by M. Deleuze at the Grand-'Combe is the easiest to mine, and that it has reserves to match any in Europe'.[9] And these reserves were not particularly difficult to mine: by the mid-nineteenth century, *without* any great technological revolution, the mines in the Alès basin would be producing not hundreds, but hundreds of thousands of tonnes of coal. The problem, throughout the period we shall be discussing, was to be supply, not demand. Since the middle ages, coal had been produced for raw-silk manufacture, lime-burners and distillers, but on a very small and localized basis. The situation had begun to change as a result of the boom in the textile industry, particularly silk, during the second third of the eighteenth century. A sign of this increased prosperity, in many parts of France, and its relationship to the development of coal-mining was the first modern Mining Act of 1744, which was an attempt (abortive in Languedoc) to introduce more modern techniques of mining by increasing the control of the state. During the second half of the century, municipalities and the Estates of Languedoc began to issue warnings on the rapidly dwindling, and increasingly expensive, supplies of wood. One recent study has rightly emphasized the importance of poor fuel supplies as one of the major obstacles to industrialization in Languedoc, something that must apply to other regions of France, like Brittany, which also had a booming textile industry in the eighteenth century.[10]

Like Brittany and Normandy, lower Languedoc failed to realize the promise of its mid-eighteenth-century 'industrial revolution'.

> By 1789, the golden age of expansion was over, the result of an economic crisis which had struck over a decade earlier ... In 1778 the commercial policy of the Spanish government led to a prohibition on stockings and silk fabrics ... In 1786, English competition, freed by the Franco-British commercial treaty, succeeded in ruining trade with the Levant, trade which had been stagnating for a long time.[11]

It is extremely important to note that moves to modernize the coal-mining industry in the Alès basin occurred within the context of a stagnating economy. This was bound to prompt the withdrawal of many investors and village communities into their traditional shell.

---

[9] *Bibliothèque de Nîmes*, Mss 469, Louis Alles to Richard (13 March 1780).
[10] See, for example E. Allen, 'Deforestation and fuel crisis in pre-revolutionary Languedoc', *French Historical Studies*, 13 (1984).
[11] Laurent and Gavignaud, *La Révolution française*, p. 15.

Serious economic recession was little more than a small cloud on the horizon of this market-orientated, merchant-capitalist and traditional manufacturing region of France when, in 1770, one of the most remarkable and entrepreneurial figures in eighteenth-century French economic history, Pierre-François Tubeuf, accompanied by his young wife, *née* Marie-Margueritte Brochet, made his first foray into the mining valleys of the Cévennes.[12] A few years later, the town of Alès would dress itself up in its best *ancien régime* finery to celebrate the visit of one of the most powerful nobles in France, Charles-Gabriel-Eugène La Croix Castries, Minister of the Navy during the American War of Independence, *confidant* of Marie Antoinette, responsible along with that irresponsible and tragic young queen for some of the most lamentable decisions which helped to precipitate the revolutionary crisis of 1787–89. In 1783, Castries would be elevated to the highest rung of the Bourbon military ladder – a *maréchal de France*.[13] For most of the 1770s, Pierre-François Tubeuf was busy laying the foundations of a modern coal-mining industry in the Alés region; for most of the following decade, the *maréchal de Castries* would lead the resistance of a proto-industrial society experiencing an economic recession against the 'monopolistic', capitalist plans of Tubeuf. The personal and legal battle between the two men, which lasted for over a decade, throws considerable light upon the conflict between 'nobles' and 'bourgeois' in the eighteenth century, enabling us to differentiate between the disruptive forces of modern industrial capitalism and the more accommodating measures taken by *seigneurs*, anxious to increase their revenues, but enmeshed, socially and politically, whether at Versailles or in the *Etats du Languedoc*, in a pre-industrial, aristocratic socio-economy. If, as Maxine Berg insists, we must give more weight to community and cultural factors when discussing the evolution of industrialization, then the weight of a 'seigneurial' culture in France before the Revolution merits particular analysis.[14]

---

[12] The *Journals* covering the activities of the Tubeuf family, which, after a very long search, I eventually tracked down in the offices of the *Houillères des Cévennes* and which are now deposited in the *Archives du départmentales du Gard*, 58 J 1–12, contain detailed accounts of the father's work in Languedoc, down to the house-keeping expenses of Madame Tubeuf.

[13] For all his subsequent counter-revolutionary activity, the maréchal de Castries was a successful Minister of the Navy, introducing many lasting reforms, collectively known as the *Code Castries*. See duc de Castries, *Papiers de famille* (Paris, 1977), p. 152.

[14] In her *Age of Manufactures*, p. 316, Maxine Berg attacks the 'rigid' models of both Marx and Franklin Mendels in favour of more complicated paths to nineteenth-century industrialization, insisting upon the necessity of including 'culture and community in our explanation of industrial and technical change'.

Constraints of Proto-industrial Society

Pierre-François Tubeuf was a man who thought that 'big was beautiful', a Landes-type entrepreneur.[15] He was a Norman by birth, and very conscious of his middle-class status. In his journals he referred to himself as 'Le sieur Tubeuf, Bourgeois'. In 1765 he had married Marie-Margueritte Brochet, daughter of a Lyon merchant. The couple were to have three children, two boys and a girl who died in infancy. Pierre-Françoise was a man of boundless energy and considerable vision; his character was marred, however, by a marked lack of tact and caution, the latter defect being the almost inevitable consequence of his entrepreneurial dynamism. His journals provide us with an intimate picture of a late eighteenth-century capitalist bourgeois. Entries include details of the purchase of Rousseau's *Confessions* and subscriptions to the *Mercure français* and the *Journal politique*, as well as the record of visits to government officials, money-lenders, and industrial pioneers like the Périer brothers at their le Chaillot works. Tubeuf was a man of some culture, who was as at home in the corridors of power as he was down the many pits he opened up in Normandy, Languedoc and Paris.

More relevant to his activities as an eighteenth-century entrepreneur are references to a M. La Barberie, first secretary to the French *contrôleur-général* in the early 1780s. La Barberie was to be the government official in charge of the *Conseil des mines* at Versailles throughout the 1770s and 1780s. He was to become a godfather to Tubeuf's elder son and a secret shareholder in Tubeuf's mining ventures. One of the most important tasks confronting entrepreneurs in eighteenth-century France was that of cultivating the relevant government civil servants and, from the beginning, Tubeuf was to prove extremely assiduous in this regard. It is the relationship of entrepreneurs with the state that distinguishes pre-revolutionary capitalism in France from that which was to emerge in the 1790s, before Napoleon re-enforced the authority of the state. One of the reasons for the economic collapse of the 1790s was to be the uncertain relationship between bankers, industrialists, manufacturers and government officials.

As a result of his constant exertions, Tubeuf obtained his original concession from the government in April 1773, which included all the coal-mines within a circle of five *lieues* around the town of Barjac. The

---

[15] David Landes, in his *The Unbound Prometheus. Technical Change and Industrial Development in Western Europe from 1750 to the Present* (Cambridge, 1969), stresses the importance of entrepreneurial figures like Tubeuf, who realized that they were living through 'revolutionary' times.

money to finance this huge concession came, in the main, from the 100,000 *livres* compensation awarded him by the government for losses sustained in coal-mining ventures near Avignon. The 1770s were to prove the decade when modern capitalist techniques were first introduced to the coal-mining industry in the south-east of France, challenging the more traditional 'proto-industrial' patterns of mining which fitted less painfully into the socio-economic and cultural realities of the region.

Pierre-François Tubeuf and the structures of modern industrial capitalism

Tubeuf's challenge to traditional forms of coal-mining was four-fold.

Scale of capital investment

The introduction of 'modern' methods of coal-mining was extremely costly, way beyond the very limited capital resources of the petty *seigneurs*, tenant farmers, small textile manufacturers, and distillers of the Cévennes. It was far more attractive, and socially desirable, to invest in land, including the cultivation of the vine as well as mulberry leaves for the production of raw silk. The majority of the sixty or so small mines in the Alès basin ran on shoestring budgets of a few thousand *livres*. According to Tubeuf's own calculations, entered in his journals, he invested no less than 400,000 *livres* in his coal-mining ventures between 1773 and 1784, and 700,000 *livres* between 1773 and the outbreak of the Revolution. Even admitting that Tubeuf was not averse to some 'creative accounting', it is clear that investment in his mines ran into several hundred thousand *livres*. We can check this from the repayments he made to his chief financial backer, the Marquis de Chaulleu, repayments which continued to be made by his wife and sons long after Tubeuf's death.[16] The scale of this investment was unparalleled in the long history of coal-mining in the south-east of France.

---

[16] Following the death of the marquis de Chaulieu and the voyage of Tubeuf to America in 1791, the marquise de Chaulieu and Madame Tubeuf decided (on whose initiative?) to share an apartment together in rue Saint-Paul, which was situated in the Marais district of Paris. By any standards it was an astonishing *maison à deux*, these two ladies forced together in revolutionary Paris by debt, death and the pull of the New World!

Concentration of production and technological innovation

When Tubeuf arrived in the region he found scores of small levels pock-marking the hills around Alès as well as the scars of many more, some dating back to medieval times.[17] Long before the eighteenth century, mines had been leased out to coal-mining families – there were 'dynasties' like the Larguiers and the Laupies in the Alès region – who produced just enough, at particular seasons of the year, to satisfy the demands of local lime-burners, distillers, forge-masters and silk manufacturers. Tubeuf's plans threatened the eventual destruction of this traditional, seasonal, medieval form of production. He closed down the great majority of small, inefficient mines, usually by buying out the leases or compensating the owners, and concentrated production on three pits: Bannes, to supply the Vivarais region; Robiac, for the Meyrannes valley; and Rochebelle, on the outskirts of Alès, to supply the town and its immediate vicinity. It is important to note that throughout this pioneering period he sought, and usually received, the support of the government. In October 1774, he thanked his friend La Barberie in the *Conseil des Mines*, adding, 'I have only one request left to make – your support for the closure of the small, unproductive mines in this region, whose owners might, by accident, inexperience, or malice, create problems for me'. This support was forthcoming, at least until the early 1780s.[18]

Tubeuf undertook technological innovation: he poured thousands of *livres* into the construction of deeper mines; his men blasted roadways through hard rock to follow, or rediscover, the main seams of coal; they dug drainage ditches for water and air-holes for ventilation. On the eve of the Revolution, he was ordering steam-pumps from the Périer brothers at their Chaillot works. Achille Bardon, a mining engineer, whose book on the Alès coal-basin was written almost a century ago, concluded that Tubeuf's work marked 'a new era in the history of coalmining in the south of France'.[19]

Organization of a new workforce

Marx paid insufficient attention to the problem of the transformation of peasants and artisans into a reservoir of wage-earning labour. Chris

---

[17] For the medieval background to the mines of the Alès basin see A. Bardon, *L'Exploitation du bassin houiller d'Alais sous l'ancien régime* (Nîmes, 1898), pp. 5–6.
[18] Bardon, *L'Exploitation du bassin houiller*, p. 84.
[19] Ibid., p. 347.

Johnson has done some pioneering work on textile workers in the nearby Hérault department.[20] Tubeuf had considerable disdain for local workers, who were far too lazy and insubordinate for him, with the result that the main cadres of his workforce were recruited from areas which had a tradition of coal-mining; skilled workers from Belgium and Germany; less skilled colliers from Piedmont. However, Tubeuf did offer employment to the miners who had been obliged to close down their own small pits. Professor Clarkson's four-pronged model of proto-industrialization includes the feature that 'industrial products were made by peasant-manufacturers who combined, say, weaving or stocking-knitting with farming'.[21] Many of Tubeuf's workers had combined digging for coal with farming or the production of raw and spun silk. This tradition of seasonal work was to create continual headaches for Tubeuf, who attempted to deal with it by the introduction of stringent regulations. His overseers (*maître-mineurs*) were instructed to be 'juste et sévère' with those beneath them. A fine of six *livres* was levied for drunkenness. Day-shifts started at 3.45 a.m. and anyone arriving late lost pay. It was forbidden for any worker to play cards underground, to sleep, smoke or defecate. Resistance to the new regime was fierce in the Cévennes. The most serious pre-revolutionary spasm of violence in the south of France occurred in 1783 when, because of a complex series of problems in which starvation and under-employment loomed large, bands of local inhabitants, known as *Masques armés*, attacked judicial officials in towns and villages in the Vivarais. It is noticeable that the worst affected regions sheltered communities which had been forced by Tubeuf to close down their mines.[22] The *Masques armés* revolt should not be viewed as a traditional peasant *jacquerie* but as a reaction against the encroachment of modern capitalism. The government took due note of this fierce spasm of resistance and Tubeuf's fortunes declined in consequence.

---

[20] See, for example, 'Patterns of proletarianization: Parisian tailors and Lodève woolens workers', in J. Merriman (ed.), *Consciousness and Class Experience in Nineteenth-Century Europe* (New York, 1979).

[21] Clarkson, *Proto-industrialisation*, p. 15.

[22] It was on 30 November 1782 that Tubeuf learned that a royal commission had confirmed the legitimacy of his original concession. On 1 December, he arrived back in Alès with his family after a lengthy spell in Paris, 'determined to execute the decrees of the *Conseil* and to stay until he had got all his money back'. ADG 58 J 2, entry in Tubeuf's Journal for 30 November 1782. For a recent analysis of the significance of the *Masques armés* revolt which took place in February 1783 see G. Sabatier, 'De la révolte du Roure (1670) aux Masques armés (1783): la mutation du phénomène contestataire en Vivarais', in J. Nicolas, *Mouvements populaires et conscience sociale, XVIe–XIXe siècles* (Paris, 1985).

Marketing of the coal

From the beginning Tubeuf's entrepreneurial vision lifted his sights far beyond a local proto-industrial society. A second feature identified by Professor Clarkson as typical of the latter is that 'craftsmen produced goods for markets beyond the regions where they lived; often these markets were overseas'.[23] Certainly this was not true of the coal producers of the Cévennes before Tubeuf's arrival, although coal was sent to Lunel, Montpellier and Nîmes. It was Tubeuf who first began to think of supplying customers in Paris and overseas. As early as 21 April 1771, he had noted in his journal: 'Returned from Marseille where I have been looking for new markets'. His declared aim was to challenge the strong competition of British coal imports in Marseille and Bordeaux, if not actually to carry 'coals to Newcastle'. His boldest scheme, however, was the contract he signed with one of the leading coking companies in France, Richard, Carrouge et Cie, to produce no less than 16,000 tonnes of coal per year for twenty-four years.[24]

It was this 'monopolist' contract that provoked powerful opposition to his ambitious plans for the region. Small property-owners, including local *seigneurs* like the *seigneur de Trouilhas*, began to be seriously worried about the income they received from leasing out their coal-mines. Coal-miners, lime-burners, forge-masters, distillers and producers of raw silk expressed concern over the availability as well as the cost of their own supplies. This was the classic case of a proto-industrial socio-economy blocking the advance of more modern forms of industrial capitalism. The opposition to Tubeuf began to pressurize the government over the huge concession awarded to this 'Norman stranger'. Did it take precedence over the rights of property-owners in Languedoc, a province governed, according to its leading spokesman, by Roman law? The complaints of the local inhabitants geared to the rhythms of a traditional economy were exploited by politicians and lawyers keen to defend the province's 'autonomy' from 'despotic' government. And who better to articulate the complaints of lawyers and lime-burners alike than Charles-Eugène-Gabriel Lacroix Castries, Ministries of the Navy, *maréchal de France*, *seigneur* of the Alès region?

---

[23] Clarkson, *Proto-industrialisation*, p. 15.
[24] ADG 58 J 2, entry for 31 July 1788 which notes that he had 'sold' 400,000 *quintals* of coal a year, for twenty-four years, to a Compagnie d'Epurement de Paris. Tubeuf added that the company had given his wife a 'sweetener' (*pot de vin*) of 3,000 *livres*. Even in the eighteenth century, business was business!

On 17 March 1777, the *maréchal* had purchased the *comté d'Alais* for 770,000 *livres*, ironically almost the exact sum that Tubeuf would invest in his coal-mines. Castries's interest in this old and prestigious fief had been fired by one of his friends at Court, the duc de Croy, who had invested in what were proving to be the richest coal-mines in France, those of Anzin. Historians like Guy Chaussinand-Nogaret are quite right to stress the involvement of many nobles in capitalist ventures, ancient and modern, and Castries was no exception.[25] In 1780, he purchased the *seigneurie* of Trouilhas, which included the richest prize in the Alès coal-field, the mine of the Grand'Combe. Battle-lines had been formed. On the one side, Pierre-François Tubeuf, armed with his concession from the king and his contract from Richard and Carrouge; on the other, the maréchal de Castries, armed with legal suzerainty over fifty parishes in and around Alès and social suzerainty over the local population. He could also rely on massive firepower at Court. The immediate question is: how did Castries approach the problem of exploiting those coal-mines which he insisted lay under his control? (Castries argued that the Grand'Combe mine, for example, lay outside, not inside, the circle of Tubeuf's original concession. To 'prove' this he suggested measurements should be taken according to the local *lieues de Languedoc*, not the national *lieues de France*!) Was this an example of noble capitalism on a modern scale? Far from it. To underline this point, let us examine Castries's approach to the four characteristics chosen to illustrate Tubeuf's attempt to modernize the Alès coal-field.

Charles-Eugène-Gabriel Lacroix Castries and the structures of a proto- industrial socio-economy

Scale of capital investment

This proved to be totally incommensurate with the money Tubeuf poured into the improvement of the mines that even Castries agreed might belong to Tubeuf, despite the fact that total income from Castries's estates doubled between 1778 and 1789.[26] There was, at least in the south-east of France, a 'seigneurial reaction' in the decades before the Revolution which obliged Castries and his estate-managers

---

[25] G. Chaussinand-Nogaret, *The French Nobility in the Eighteenth Century: from Feudalism to Enlightenment* (Cambridge, 1985), chapter 5.

[26] Details of the maréchal's accounts are contained in *Papiers de famille*, pp. 315–19.

to be even more protective of the rights of those property-owners paying the increased dues, including those who earned part of their own income from leasing out coal-mines. One of the major issues which tied nobles like Castries into the old structures of a rural and aristocratic economy, and which explain the uneasy alliance between Castries and his vassals, was the problem capitalists like Tubeuf posed by closing down coal-mines. If *seigneurs* had lost the military duty of defending local communities, their social duty of defending them against 'monopolists' and 'strangers' like Tubeuf persisted. And this is the key issue. In the final analysis, Castries depended upon his stewards and seigneurial judges for the implementation of his plans to increase his own revenues; they were the essential brokers of the seigneurial system, bridging the gap between great *seigneurs* like Castries and the poorest collier or lime-burner in the locality. These seigneurial brokers made their own living from the administration of noble estates. Thence came the bitterness of their opposition to outsiders like Tubeuf, who threatened the old ways of doing things. I will give one telling example.

In 1785, Castries received advice from one of his agents that it might be profitable, in the long term, to construct a forge near the Grand-'Combe mine. The suggestion was immediately rejected by the *maréchal*'s steward in Alès, Crozade, who wrote: 'Individual undertakings such as these are only suitable for individuals who can manage them directly, and who can devote all their energies to finding markets for the products they manufacture. We are talking of industrial ventures which rarely succeed in the hands of great nobles like you, because these kind of ventures inevitably involve great expense and little profit.'[27]

Concentration of production and technological innovation

Nothing was done to concentrate production in key areas: Tubeuf had already 'cornered the market' in the Vivarais and in Alès itself. Castries's aim, or rather that of his stewards, was to continue to satisfy local suppliers, particularly those forge-masters, distillers and silk producers who traditionally relied on the good-quality coal of the Grand'Combe. As for technological change, little attempt was made to imitate Tubeuf's pioneering innovations at Bannes and Rochebelle. What Castries wanted was quick profits, so that the medieval system of ripping out the coal within easiest reach was continued. In 1790, the

[27] AN 306 AP 487, Crozade to Castries, 25 June 1786.

main seam in the Grand'Combe mine had become unworkable owing to roof-falls and flooding.[28]

Organization of a new workforce

Until 1785, Castries leased out his mines to a *société des fermiers de la Grand'Combe*, a group of men centrally involved in organizing the fierce opposition to Tubeuf while persisting with the traditional methods of financing and production. Eventually, after allegations that they had been swindling the *maréchal*, Castries's steward turned to a German *maître-mineur*, Adam Forster, who had originally worked on Tubeuf's concession. Attempts to lure away Tubeuf's main mining engineer Renaux, a graduate of the recently formed *Ecole des mines* in Paris, were blocked by Castries's steward, who regarded a man who was to prove the most knowledgeable and reliable mining engineer in the south-east as 'too young and inexperienced'.[29] Most of Castries's miners had worked for the corrupt and totally inefficient *société des fermiers de la Grand'Combe*.

Marketing of the coal

Coal from the Grand'Combe was bought by the old local clientele from the region. When Castries received a personal request from a friend in the *Etats du Languedoc* for a contract to supply 1,000 tonnes of coal a year, Crozade warned that it would be better to refuse. He explained that allowing that much coal to leave the immediate Alès region would produce a 'murmure général'. In other words, Crozade was afraid of a repetition of something on the scale of the *Masques armés* revolt of 1783. It has to be remembered that stewards like Crozade lived and worked among the local population, and were obviously marked by the imprint of a traditional economy and society. Castries's job was to protect, not to disrupt it.

There can be no doubt that, to understand the 'transition from feudalism to capitalism' we must go beyond the rigid framework of marxist and proto-industrial models 'to include culture and community in our explanations of industrial and technical change'.[30] However,

---

[28] In a letter to his *directeur des mines*, Renaux, dated 24 May 1790, Tubeuf asks for details of the *catastrophe* which had befallen the Grand' Combe mine. Bardon, *L'Exploitation du bassin houiller*, p. 335.

[29] ADG 58 J 2, Crozade to Castries, 24 September 1786.

[30] Berg, *Age of Manufacture*, p. 316.

the debate on proto-industrialization has undoubtedly opened up new avenues of enquiry, leading, in some instances, to that desirable association between history, economic history and social history which Dr Berg advocates. An example of this is the third characteristic of proto-industrialization mentioned by Professor Clarkson: 'that rural manufacturing stimulated commercial farming by creating a market for food for their own needs'.[31] Again, Tubeuf, through his ill-fated contract with Richard, Carrouge et Cie in particular, threatened to disrupt this delicate relationship between the manufacturing and farming regions of the south-east.

The economy of the Cévennes region was notable for the close relationship which existed between coal and bread. It was estimated that, in the 1780s, 1,391 carts per annum from the plain around Nîmes brought in grain, salt, and other basic commodities, returning laden with 2,000 tonnes of coal.[32] The monopolistic designs of Tubeuf threatened this traditional system of exchange, upon which entire communities depended for their existence. No coal leaving the region meant no grain coming in, and that evoked the fears of starvation and death that were part of the collective cultural DNA of most eighteenth-century societies. And who controlled the grain trade? Often the very same stewards and seigneurial judges, or their close acquaintances, who were in the vanguard of the movement to defeat Tubeuf. As Cabane de Camont, seigneurial judge for the maréchal de Castries told the secretary to the *Etats du Languedoc*: 'Many people in Alès fear, not without reason, that the exclusive concession granted by the king to sieur Tubeuf, together with the contract the latter has signed with Richard, Carrouge et Cie for the extraction of coal, may well become an exclusive privilege for the sale of grain'.[33]

Coal and bread were inextricably connected in the minds of eighteenth-century Alèsiens. Given that the economic recession of the 1780s had increased fears of mass unemployment and starvation, it can easily be appreciated why Tubeuf's plans provoked such fierce controversy. 'Elites' and 'the people' both stood to lose out. The forces of modern industrial capitalism were faced with the traditional barriers of social and economic interdependency. The *Masques armés* revolt had provided a warning of what might happen if the local elites failed to protect the interests of the people.

Through the maréchal de Castries, local elites found the perfect

---

[31] Clarkson, *Proto-industrialisation*, p. 16.
[32] *Bibliothèque de Nîmes*, Mss 496, Alles to Richard, 19 June 1780.
[33] ADG C 194, Cabane de Camont to Alexis Trinquelaque, 21 January 1779.

instrument to defeat Tubeuf's admittedly over-ambitious plans. Castries was a powerful member of the *Etats du Languedoc*; he presided over the meetings of its subsidiary body, the *assiette du diocèse d'Alais*, at which resistance to Tubeuf was first organized on a major political level. From Alès and Montpellier, Castries could take the issue literally to the seat of the throne at Versailles. Convinced, rightly from his standpoint, that on this occasion his interests and those of the local Alèsiens were identical, the *maréchal* took Tubeuf to court, and to Court, on several occasions during the 1780s, forcing Tubeuf to move the centre of his operations from Languedoc to Normandy, and then to Paris.

On the very eve of the Revolution Castries, by packing the extraordinary royal council convened to make a final decision on the Tubeuf/Castries affair, won a legal victory which virtually ruined the Norman entrepreneur. Tubeuf, despite the concession from the king, was deprived of his rights to the Grand'Combe, as well as one or two other mines, and ordered to pay Castries a quarter of a million *livres* indemnity. The *maréchal*'s descendant, the historian the duc de Castries, writes in his *Papiers de famille*: 'This legal decision is most interesting since it confirmed, six months before the meeting of the Estates General, that seigneurial rights prevailed over all others, whether or not they were granted by a king.'[34] There was indeed a 'feudal' reaction in the 1780s, and that reaction operated at Versailles as it did in the seigneurial fields of *ancien régime* France.

The maréchal de Castries was to become one of the leaders of the counter-revolution in the 1790s; Pierre-François Tubeuf crossed the Atlantic to Virginia, where he prospected for iron and coal in an environment more conducive to modern capitalism than that of eighteenth-century France, at least as far as the white colonists were concerned. In his journal, Tubeuf expresses his conviction that, in the final analysis, his defeat was due to the superior political and social clout of the maréchal de Castries; interestingly enough, a *maréchal* who was far too busy at Court to deal with the complexities of the situation in the Alès coal-field. The immediate cause of his defeat, according to Tubeuf, was the *maréchal*'s agents and advisers in the south-east. He also blamed the government for failing to give him the support he had enjoyed in the early years of his concession. Just before the meeting of the Estates General was to transform France for ever, one government official had criticized Tubeuf for being 'too enterprising'. Tubeuf's reply is revealing: 'You accuse me of being too

---

[34] Duc de Castries, *Papiers de famille*, pp. 306–7.

enterprising (*trop entreprenant*). I cannot accept this criticism since I have only involved myself in ventures which have been useful to the State, having heard a Minister say out loud that the State would be happy indeed to find individuals like me who devoted themselves to the public good with the same zeal and the same happy results that I have.'[35] In this reply, we see revealed the struggle between government officials, like La Barberie, who had backed Tubeuf to the hilt, and those who were too enmeshed in the aristocratic web of intrigue and personal advancement at Versailles to concern themselves with the wider (national) public interest. One of the greatest constraints operating upon the advance of industrial capitalism in France was a political and social schizophrenia on the part of the government of Louis XVI, something which was, surely, to be one of the major causes of the Revolution.

This 'schizophrenia' between the demands of modern capitalism and those of a traditional, landed society is also central to an understanding of the Revolution. I have dealt with this at length in a book that traces the issues raised in this essay up to the Revolution of 1830.[36] As far as the Revolution of 1789 is concerned, we may simply note here that the 1790s were to prove a disastrous period for the coal-mines of the Alès basin. Those historians with more ideological than historical sense might dismiss this fact as fundamentally attributable to the disruption caused by the Revolution itself. Certainly, the strains imposed by repeated political and social upheavals, as well as the exigencies of war which involved the *immediate*, as opposed to the planned, long-term production of coal, were instrumental in explaining the fact that by the time Napoleon came to power in 1799 the output of the Alès coal-basin was significantly lower than the 15,000 tonnes recorded for 1789.

One has to recall, however, that Pierre-François Tubeuf was planning, *ten years before the Revolution*, to supply Richard, Carrouge et Cie alone with 16,000 tonnes of coal a year. The stagnating, indeed falling, levels of production recorded for the Alès coal-field during the last decades of the eighteenth century and the first two decades of the nineteenth century cannot possibly be explained without reference to the lack of anything approaching the capital investment supplied by Tubeuf a generation earlier, as well as to the continued resistance of

---

[35] ADG 58 J 5, Tubeuf to Milliner, 8 April 1789.
[36] G. Lewis, *Maréchals and Mine-owners: Proto-industrialization, Revolutions and the Retardation of Modern Capitalism in the Basses-Cévennes (1770–1840)* (Oxford, forthcoming). Chapters 3 and 4 cover the period of the Revolution and Empire.

'proto-industrial' communities pushed further in the direction of 'safe' investments in sericulture and viticulture by the uncertainties of war and revolution. In many important spheres of policy, the French Revolution served to protect the old, rather than to welcome in the new. This was to be particularly true of the rights of property-owners.

In March 1791, Tubeuf distributed among the deputies of the Constituent Assembly 900 copies of a *mémoire* on mining in which he explained why he believed the rights of entrepreneurs like himself should take precedence over those of individual property-owners.[37] Its impact was limited. The legislation governing mining in France, passed by the Assembly on 28 July 1791, was a scarcely qualified victory for property-owners, who were given priority in deciding the fate of any coal lying beneath their land. The position would only be reversed in favour of the state, through the provisions favouring entrepreneurs who possessed the necessary capital and technical expertise, towards the end of the Napoleonic Empire (April 1810). Patrice Higgonet is absolutely right to emphasize the fact that, 'Epistemologically and politically ... the contradictory claims of community and individualism had never been satisfied in the eighteenth century.'[38] The Revolution was to be the offspring of this unconsummated marriage, as the legislation of 28 July 1791 illustrated. On this date, Pierre-François Tubeuf was nearing the end of an epic seventy-four day crossing of the Atlantic, on his way, with his eldest son, to start a new life in what he considered to be the ideologically unfettered world of Virginia, USA. In what must surely be one of the great ironies of history, he was to be murdered, while prospecting for iron ore and coal, in 'the back country', by representatives of another community defending its traditional rights and culture – '*les peaux rouges*'![39]

---

[37] ADG 58 J 5, 25 March 1791. Entry recording the distribution of 900 copies of his *Idées générales sur les mines de charbon de terre, sur les concessions*.

[38] P. Higonnet, *Class, Ideology, and the Rights of Nobles during the French Revolution* (Oxford, 1981), p. 16.

[39] ADG 58 J 8, letter of William Alexander, n.d., recording the 'brutal massacre' of Pierre-François Tubeuf 'in the Back Country where your family possesses a large and valuable property'.

5
Technology and Innovation in an Industrial Late-comer: Italy in the Nineteenth Century

John A. Davis

Introduction

What part did the failure to adopt new technologies and processes of innovation play in retarding economic growth in a typically slow mover in the early phases of the industrial revolutions like Italy? The need to adopt and adapt to the new technologies that were being developed elsewhere was certainly perceived to be a major problem by contemporaries, and from the mid-eighteenth century onwards there were many Italians who saw the persistence of archaic techniques and methods of production as a primary reason for the failure to emulate the industrial expansion pioneered by the economies of north-western Europe. Both individuals and the governments of the separate Italian states – before political unification in the mid-nineteenth century – devoted great energy and expense to studying new methods and technologies that were being developed abroad.

Yet despite numerous initiatives in this direction, the results were generally considered to be disappointing. Indeed, it was not until the very end of the nineteenth century, when industrial growth in Italy began to take on greater vitality and breadth, that the sense of technological backwardness began to disappear. But even then, although Italy, like many other late industrializers, was able to take advantage of the new technologies of the 'second industrial revolution' – most notably hydro-electric power, advanced processes for steel production, chemicals and synthetic materials, as well as new materials like rubber and advanced engineering products such as typewriters (Olivetti) and motor cars (FIAT and ALFA) – economic historians have pointed to the continuing technological shortcomings of Italian industry and its reliance on a group of low-technology sectors, especially textiles, until the First World War.[1]

To suggest, however, that there was a technological barrier which impeded the development of manufacturing and industry in Italy for over a century would be a gross distortion. To start with, the failure to innovate was never absolute. As well as some major technological breakthroughs, more constant processes of technological change were taking place in a variety of different sectors of the economy over relatively long periods – although not always ones that led on to wider processes of innovation and growth.

Even more important, Italy had by the time of the First World War become an industrial power. In 1918 she was officially ranked as the eighth industrial power in the world – sixth for steel production, fifth for cement, sulphuric acid, automobiles and hydro-electric power, fourth for superphosphates and artificial textile fibres.[2] In view of the overwhelming weight of a largely unproductive and poor agrarian economy, not only at the close of the eighteenth century but right up to the last decades of the nineteenth century, this achievement was remarkable and constitutes a fundamental corrective to any analysis that fastens solely on perceived technological shortcomings in Italy's industrial structures. It should also be remembered that none of Italy's southern European neighbours, not even Spain despite its much richer natural resources, had reached comparable levels of industrial growth before 1914.

Italy's economic history in the period from the mid-eighteenth century to the First World War cannot, therefore, be seen as one of unrelieved technological backwardness and immobility. But the view from this late-comer does provide important perspectives on the more general questions of how technology transfers occurred as well as on

---

[1] The best descriptions in English of the Italian economy in this period are: L. Cafagna 'The industrial revolution in Italy 1830–1914', in C. Cipolla (ed.), *The Fontana Economic History of Europe*, vol. 4 (London, 1973) and G. Toniolo, *An Economic History of Liberal Italy 1850–1918* (London, 1990).

[2] G. Mori, 'The process of industrialization in general and the process of industrialization in Italy', *Journal of European Economic History*, 8 (1979), p. 64.

the relationship between technological innovation and economic growth. In particular, the Italian case leads us to ask to what extent technological backwardness initially impeded economic growth, and then how and why these technological barriers were broken down. In both cases, the Italian experience suggests that, notwithstanding the critical importance of the technological factor in economic growth, technology alone was never the determinant factor in either retarding or accelerating the pace of growth. On the contrary, the patterns of technological development that occurred in Italy in this period were more often dictated by other factors of production. This in turn would suggest that for an industrial late-starter, the problems posed by technology transfer and innovation were not always as straightforward as some economic historians would have us believe.[3]

Technology and economic growth before unification

The eighteenth century

Italian observers were increasingly aware of the relative technological backwardness of their economies from early in the eighteenth century. As early as 1719, for example, Andrea Tron, a Venetian nobleman and ambassador at the court of St James, brought the first flying shuttle to Italy for use in his woollen mill near Treviso, where he also adapted an English process for mechanical teaseling.[4] But it is also true that contemporaries well understood that economic growth did not depend on manufacturing and new technologies alone. Rightly, they were no less concerned by the constraints on economic growth imposed by an agrarian economy that was in many cases backward and unproductive, dominated in much of the peninsula by subsistence peasant farms that relied on low levels of productivity that could not sustain increasing commercial pressures. The situation was aggravated in many parts of Italy by the persistence of feudal restrictions on the ownership and exploitation of land and by the existence of seigneurial monopolies over other activities. No less important was the persistence of mercan-

---

[3] See S. Pollard, *Peaceful Conquest* (Oxford, 1981), pp. 85–121; A. Milward and S. Saul, *The Economic Development of Continental Europe*, Vol. 1 (London, 1973), pp. 30–40; C. Trebilcock, *The Industrialization of the European Powers* (London, 1981), pp. 129–30.

[4] S. Ciriacono, 'L'industria a domicilio nel Veneto dell'Ottocento', in A. Lazzarini (ed.), *Trasformazioni economiche e sociali nel Veneto fra XIX e XX secolo* (Vicenza, 1984), p. 583.

tilist policies and guild controls over many branches of urban manufacturing.

These were rightly seen as primary obstacles to economic growth and the most fundamental aspects of the Enlightenment debate on economic development in the Italian states were concerned with finding ways to regenerate agriculture and commerce in general. None the less, technological development attracted growing attention, but was rarely seen solely in terms of manufacturing alone. As eighteenth-century travellers noted, it was often in agriculture that the most important technical developments occurred. An important example was land reclamation, a major factor in economic growth given the vast areas of uncultivated land in Italy. Another was irrigation, and when Arthur Young travelled through the Milanese plain towards Lodi in 1789 he was awed by the extent and complexity of the irrigation systems he found there and the richness of the meadows and dairy farms they supported. Further west, on the plains around Novara and Vercelli, major investments were being made in the same period in the development of rice cultivation, which led to huge increases in productivity. Not all the Po Valley had achieved the same level of development, of course, and towards the east it was not until the late nineteenth century that technological progress made it possible for immense areas of swampland to be reclaimed for cultivation. Throughout central and southern Italy conditions were generally much poorer, yet with notable exceptions; even in the most remote regions eighteenth-century travellers found examples of improving land-owners who were experimenting with new crops, new rotations and other innovations from abroad.[5]

Even allowing for the overwhelming predominance of agriculture as the primary sector of industrial activity in eighteenth-century Italy, there was still an immense variety of manufacturing activities and industries. Many – the majority – were closely linked to agriculture and catered mainly for the needs of the producers and their families; but there were other sectors which were fully integrated in production for international markets.

If the economy of the *ancien régime* Italian states had declined sadly since the glories of the Renaissance, Italy was by no means a technological desert at the close of the eighteenth century. Yet the changes that were taking place in the structure of the international economy were also beginning to have profound effects on the economies of the Italian states. The expansion in the volumes of international

---

[5] For example, Arthur Young, *Travels in France and Italy* (London, 1915).

trade, the shift in demand away from high-cost luxury goods towards cheaper products (especially textiles, of course) and the rapid increase in the demand for primary goods for use in manufacturing processes had far-reaching consequences for manufacturing activities throughout the Italian peninsula.

The impact of these changes was particularly evident in the case of the most important and valuable manufacturing activity in Italy: the production and working of silk filaments and the weaving of silk fabrics. Silk was Italy's single most important export commodity in the eighteenth century, and a century later still provided three times more value added than the chemical, engineering and metal-making industries combined.[6] But in the intervening period the industry had undergone a complete transformation, in which the traditional silk twisting and weaving industries virtually disappeared while the production of cocoons and spun warps and wefts had increased massively.

These changes were accompanied by critical technological developments. At the beginning of the eighteenth century the processes for reeling and twisting silk that had been developed in northern Italy, notably at Bologna, were rated the most advanced in the world. It was the Bolognese spinning mill that John Lombe and his brother copied and took to Derby in the early eighteenth century to found the English silk industry in what has been called one of the most spectacular cases of industrial espionage in modern history.[7]

Carlo Poni has minutely reconstructed how as early as the seventeenth and eighteenth centuries the response to increased demand for raw silk resulted in the gradual adoption throughout northern Italy of new types of mills for reeling and spinning silk. Although the water-driven Bolognese silk twisting mills constituted the most advanced silk throwing machinery in seventeenth-century Europe, they were rapidly displaced in the course of the eighteenth century by new 'Piedmontese' mills that produced 'organzine' yarn, in which the main twist runs in a contrary direction to the strands to give the yarn greater strength.

---

[6] Toniolo, *Economic History of Liberal Italy*, p. 109; Cagagna, 'Industrial revolution in Italy', pp. 305–9; S. Fenoaltea, 'The growth of Italy's silk industry (1861–1913)', *Rivista di Storia Economica*, new series, 5 (1988), pp. 275–318.

[7] C. Poni, 'Alle origini del sistema di fabbrica: tecnologia e organizzazione produttiva dei mulini di seta nell'Italia settentrionale', *Rivista Storica Italiana* (1976), pp. 444–98. For a description of how silk is worked and processed see Fenoaltea, 'Growth of Italy's silk industry', pp. 276–8, and K.R. Greenfield, *Economics and Liberalism in the Risorgimento: a Study of Nationalism in Lombardy (1814–48)* (Baltimore, 1965), pp. 82–6.

This responded to market demand for stronger and better quality silk yarns, but adoption of the new technologies for reeling and spinning had far-reaching consequences. In Bologna, silk twisting had been carried out since the fifteenth century in what can only be described as factories, but elsewhere silk reeling and throwing was largely a domestic and decentralized activity. However, the new Piedmontese mills created strong incentives for centralizing other secondary processes and throughout the eighteenth century new reeling and throwing mills began to spring up around the abundant Alpine streams of northern Piedmont and the Veneto.

A report from the Piedmontese town of Racconigi in the early eighteenth century revealed that the new mills had displaced many hand-spinners and caused unemployment, while greatly increasing the demand for cocoons and water supplies. Later in the century there were growing complaints that the concentration of the new reeling mills around Turin had caused silk reeling in other parts of Piedmont to decline. As the number of mills in and around Turin increased, they attracted growing numbers of workers who became separated from their families. By the end of the century it was estimated that there were some 4,000 workers in the Turin silk-reeling mills who were totally dependent on the industry for their livelihood. The water-powered Piedmontese mills also reduced the need for physical strength on the part of the operatives, so that women and children began to replace male workers. In the early eighteenth century 68 per cent of the silk spinners in Piedmont were women and children and in Bologna it was customary for families to hire their children out to silk masters for periods of three years.[8]

There was no single pattern, however. In Milan the collapse of the traditional silk weaving industry was accompanied by a revival of silk reeling and saw a shift from the city back into the countryside in the late eighteenth century, with the appearance of mills producing Piedmontese-style organzine thread in rural areas with plentiful supplies of water power and close to farms where the cocoons were reared.

Many established centres of silk production proved unable to adapt to the new technology. This was the case for Bergamo, in the Veneto, which had traditionally produced highly-prized organzine yarn that was sold in Milan, Switzerland, France, England and Holland. In the course of the eighteenth century Bergamo rapidly lost its markets, partly because the growing demand for cocoons reduced its supplies of

[8] Poni, 'Alle origini del sistema di fabbrica', pp. 471–3.

raw silk and partly because Bergamo manufacturers failed to adopt automatic winders in their spinning mills, which fatally limited their productivity and increased labour costs. Even more spectacular was the case of Bologna, the pioneer of the European silk industry. Despite its technological lead in silk twisting, silk production in Bologna rapidly declined in the eighteenth century because it failed to adopt the new Piedmontese mills. The reasons for this, Carlo Poni has argued, were not technical, but had more to do with the rigidities of guild organization and the high costs of urban production, together with the fiscal and mercantilist policies of the Papal States to which Bologna was subject.[9]

The new technologies also had damaging consequences for domestic producers using older methods. This was especially evident in the Kingdom of Naples, where the government attempted to respond to the drastic crisis of the southern silk weaving industry by setting up a model silk factory at S. Leucio alongside the royal palace of Caserta. The S. Leucio silk works were modelled on the *'fabriques royales'* of *ancien régime* France, consisting of a factory and workers' colony, with workshops where the silk could be reeled and spun, and an area for weaving sheds. The colony started operating in 1776 with eighteen families and by 1788 employed 214 individuals – a number of whom were master weavers who had been recruited in Piedmont and from Lyons (despite protests by the French government), which indicates the difficulty of finding skilled workers locally.[10]

The S. Leucio enterprise quickly ran into difficulties, even though it did succeed in finding export markets for its quality silk fabrics – at least before the Revolution in France threw Mediterranean trade into chaos. But from much earlier the initiative had run into major internal problems, many of which arose from the new technologies.

In order to improve the quality of silk yarns, the S. Leucio colony was equipped with the Piedmontese spinning mules, and to encourage their diffusion the use of other equipment for reeling and twisting silk was banned throughout the Kingdom. The reason, according to a decree of 1805, was to improve the traditional methods of reeling which relied on 'the rough limbs and equipment of folk whose hands are coarsened and roughened by working with the plough and spade,

---

[9] Ciriacono, 'L'industria a domicilio nel Veneto', pp. 66–9; C. Poni, 'Per la storia del distretto industriale serico di Bologna (secoli XVI–XIX)', *Quaderni Storici*, new series, 73, xxv (1990), pp. 93–167.

[10] B. Caizzi, *Storia dell'industria italiana. Dal XVIII secolo ai nostri giorni* (Turin, 1965); M. Battaglini, *La Fabbrica del Re* (Milan, 1984); G. Tescione, *L'arte della seta a Napoli e la colonia di S. Leucio* (Naples, 1932).

causing damange to the stuff and reducing its value by more than half'. In all those areas where Piedmontese-style mills and mules had been introduced the sale of non-organzine yarn was banned and peasant producers were obliged to sell their cocoons to the new mills.

By 1806 there were eighteen Piedmontese-style mills operating in the province of Caserta and in the neighbouring province of Terra di Lavoro – although there were only two registered in the more distant province of Calabria. However, the official reports reveal that the new mills aroused great hostility among the local population, not least because the attempt to impose the new technologies risked destroying the domestic spinning industries in the areas around the new mills. Hence the particular hostility towards the most ambitious of the new ventures, the S. Leucio manufactory. The persistent desertions and thefts of raw materials, spinning frames and looms from the colony almost certainly reflected this hostility, since many of the S. Leucio workers came from the neighbouring population, whose own industry was damaged by the royal enterprise.

When in 1806 the Bourbon monarchy was deposed by Napoleon's armies, the French administrator who reported on the state of the silk industry in the Mezzogiorno did not hesitate to identify the clash between the interests of the peasant producers and those of the market as the principal obstacle to technological development. Indeed, it was this, he argued, that had made the abolition of the long-standing fiscal monopolies over silk production by the Bourbon monarchy ineffective, 'because a very numerous class totalling some 1,500 families of silk-reelers have obstructed the progress of the production of organzine yarn, which is so greatly to be desired, since this threatens to force into idleness their coarse limbs and machinery'.[11]

Similar problems would recur in other branches of textile production in the south when new technologies were introduced. During the Napoleonic period, for example, a number of cotton factories were started in the south, mainly by Swiss entrepreneurs whose trade had been destroyed by the revolutionary wars. The attempts to cultivate cotton in the south were unsuccessful, but the cotton mills survived in the region around Salerno and further north in the Liri valley. Yet they remained a constant object of popular hostility throughout the early nineteenth century, not only because of dislike of the foreign and Protestant skilled workers they employed but also because the mecha-

---

[11] *Archivio di Stato di Napoli*; Interior Ministry (II) f.5066, 29 Luglio 1805/23 maggio 1806.

nized production of cotton yarn greatly limited the scope for domestic production.[12]

As well as silk, there were numerous other textile industries that produced for inter-regional or international markets in eighteenth-century Italy. Wool was worked almost everywhere, but more specialized woollen cloth was produced in the pre-Alpine valleys of northern Piedmont around Biella and in the valleys above Vicenza in the Veneto where strong export markets were established. Flax was plentiful in the Po Valley and there were thriving linen and hemp industries. The first successful attempts to establish cotton-spinning – mainly by Swiss entrepreneurs – also date from the late eighteenth century, but increased during and after the Napoleonic period.

There were also numerous mining and metal-working industries. The island of Elba contained the richest mineral deposits in Italy and provided the ores for the metal-working industries on the Tuscan mainland at Portoferraio. There were also ironworks in Calabria, where a royal foundry was created in the eighteenth century. But the richest mineral deposits lay in the Alpine valleys, where there were numerous mining and metal-making industries that had achieved high levels of specialization and good export markets. In the early eighteenth century there were eighty-four iron mines operating in the province of Bergamo, with eleven blast furnaces and fifty-eight forges, employing in all some 3,000 to 4,000 workers. Above Brescia, the Trompia, Sabbia and Camonica valleys were the centres for highly specialized metal-working, especially the manufacture of firearms, swords and daggers.[13]

During the course of the eighteenth century nearly all of these industries entered into crisis, some virtually disappearing and others undergoing major internal transformation. There was no single cause, but the technology factor was rarely absent, and the nature of the changes taking place gave the issue of technological development increasing importance.

In the case of the Italian mining and metal-working industries, the principal constraint was the dearth of fuel other than timber, the high cost of charcoal and the total absence of coal. In some cases the lack of

---

[12] J. Davis, *Merchants, Monopolists and Contractors: Economy and Society in Bourbon Naples (1815–60)* (New York, 1981), pp. 108–32.

[13] S. Ciriacono, 'Proto-industria, lavoro a domicilio e sviluppo economico nelle campagne venete in epoca moderna', *Quaderni Storici*, 52, xviii (1983), p. 60; F. Facchini, *Alle origini di Brescia industriale* (Brescia, 1980); A. Kelikian, *Town and Country under Fascism: the Transformation of Brescia 1915–26* (Oxford, 1986), pp. 9–20.

timber simply forced mines to close, although the Mongiana foundry in Calabria moved peripatetically from site to site to eke out the available supplies of timber – a solution that hardly encouraged technological development. In Piedmont and northern Lombardy the problems posed by shortages of timber were further exacerbated by growing competition from the Austrian producers of Carinthia. Even for specialized manufactures, the challenge from new sources of production could threaten established markets. This made it all the more important to improve the quality of production and reduce costs, but the high cost of imported coal ruled out experiments in new processes of iron smelting and forging.

Given the difficulties posed by technological change an alternative response was to reduce costs. Since the mining and forging operations of the Alpine valleys were in general closely integrated with agriculture, and used labour that was part-artisan and part-peasant, the simple solution was to reduce the periods in which the forges and mines were worked – a solution that risked triggering a process of de-industrialization.

Increased competition on long-established markets was felt in other sectors too, and both public authorities and private entrepreneurs made efforts to improve the quality of production through adopting new techniques and technologies. In the 1740s, for example, a Friulian entrepreneur named Giacomo Limussio tried to halt the collapse of the Venetian linen industry by establishing a manufactory for producing mixed linen and hemp cloth together with a printing works at Tolmezzo. The initiative was imitated by many others, but failed and only in part for technological reasons. The greatest damage was caused by Habsburg customs policies which sought to restrict access to markets elsewhere in the Empire, and by the rising cost of raw materials (another sign of increased production and competition).[14]

The early nineteenth century

The structural shifts and changes that were already evident in the eighteenth century took on greater intensity in the early nineteenth century, and were once again especially evident in the silk industry. Silk production – especially in Piedmont and Lombardy – grew in importance as the peninsula's principal export commodity and came to be known as Italy's 'white coal'. But as the shift towards the rearing of

[14] Ciriacono, 'Proto-industria', pp. 65–75.

cocoons and the primary processes of reeling and throwing became even more pronounced, markets for raw and semi-worked silk also became increasingly competitive. Great Britain, for example, which had been the principal purchaser of raw and spun silk from northern Italy, found a new and cheaper source of supply in Bengal, with the result that Italian exports of raw and reeled silk to England dwindled after the 1820s. To retain their French, Swiss, Austrian and German markets, Italian producers had either to improve the quality of their semi-worked silk or to become simple suppliers of cocoons.

Production of cocoons expanded very rapidly and was concentrated in the hill regions of Piedmont and Lombardy, where agricultural leases were modified to force peasant farmers to devote more space to mulberries and cocoons. While the production of cocoons was heavily decentralized, the primary processes for winding the silk filaments from the cocoons were increasingly carried out in small specialized mills situated in the areas of production, many of which began to introduce steam-powered vats that greatly increased production and improved quality.[15]

As well as the centralization of reeling operations, a small number of silk-throwing mills were established in Lombardy in the same period, and tended to be located close to commercial centres. Whereas most reeling mills operated only for a few months in the year, production in the throwing mills was more permanent. A further incentive for developing these manufacturing processes came from the outbreak of pebrine that decimated cocoon production in the 1850s, and a number of enterprises began importing raw silk for manufacturing yarn in water-powered mills.

The slow mechanization of silk throwing brought about a further displacement of male workers by female and child labour. The gains were essentially managerial, however. Female and child labour was cheaper than male, while the centralization of production gave the manufacturers greater control over the processes of production. Anna Bull has argued that agitation among the male silk spinners in the Comasco region in the 1860s played a crucial part in encouraging the silk manufacturers to mechanize the throwing and twisting operations, which enabled them to employ cheaper and more docile female and child labour. Similar motives lay behind the moves a decade or so later to mechanize silk weaving, which again resulted in deskilling and a

---

[15] Greenfield, *Economics and Liberalism in the Risorgimento*; A. Dewerpe, *L'industrie aux champs: Essai sur la proto-industrialisation en Italie du Nord (1800–1880)* (Rome, 1985).

switch from male to female labour.[16]

This also accentuated the rural character of the Lombard silk industry, and by reducing the need for male labour the new technologies helped preserve the symbiosis of industrial production and rural labour. From the point of view of the silk manufacturers there was, of course, nothing irrational in this. Their interests and those of the land-owners had always been very close. Cheap rural labour attracted the silk manufacturers, while the presence of seasonal bye-employment helped the land-owners retain an abundant labour force without having to pay higher wages.

Similar developments occurred in the Biellese woollen industry in northern Piedmont. Domestic woollen weaving was carried out throughout the peninsula, mostly for direct or at best local consumption, but there was also a small number of regions that had established firm links with supra-regional and international markets and Biella was among the most important.

The Italian historian Franco Ramella has made a detailed study of the ways in which woollen production in the Biellese made the transition from an artisan-based proto-industry to factory-based production in the course of the nineteenth century. At the beginning of the century woollen weaving in the mountainous valleys above Biella was still closely controlled by the fiercely independent communities of weavers. The weavers were also small land-owners and their farms gave them a degree of autonomy which was reinforced by a network of tight-knit community values that enabled them to exercise effective control over access to the local labour market.

When water-powered spinning mills were introduced in the 1820s the weavers did not resist. The water-powered spinneries increased the output of yarn and provided employment for the weavers' daughters. Stern patriarchal organization kept the family unit intact so that mechanized spinning continued to co-exist with hand-loom weaving without seriously weakening the weavers' control over the local labour market. But when the employers tried to break that control this resulted in a series of bitter conflicts, culminating in one of the first extensive outbreaks of organized strike action in Italian history in the 1860s. It was this, in Ramella's view, that led the employers to invest in mechanized weaving in the following decade, even though the pur-

---

[16] A. Cento Bull, 'The Lombard silk industry in the 19th century: an industrial workforce in a rural setting', *The Italianist*, 7 (1987), pp. 99–121, and 'Proto-industrialization, small-scale capital accumulation and diffused ownership. The case of Brianza in Lombardy (1860–1950)', *Social History*, 14 (1989), pp. 177–200.

chase of the new mechanized looms greatly increased production costs and could not be justified in terms of either market opportunities or clear gains in productivity.[17]

The developments in the silk and woollen industries in northern Italy in the mid-nineteenth century suggest that technological development was driven more by managerial concerns than by concerns over productivity. Indeed, the volatility of the markets meant that the flexibility of operation offered by cheap rural labour that could easily be laid off when demand was slack was of critical importance and in many cases inhibited the introduction of more productive technologies that required more continuous cycles of production to repay investment costs.

As well as cheap and abundant labour supplies and highly unstable markets, the cost of fuel and raw materials raised obstacles to technological innovation. Combustible fuels were costly and of poor quality owing to the absence of coal and the heavy deforestation of the Alpine valleys, which continued to impede the development of mining and metal-making industries. In 1841, for example, Signore Rubini received a gold medal for the improvements introduced in his ironworks at Dongo on Lake Como, but although its furnaces were the largest in northern Italy they were still fired by charcoal. The high cost of fuel also delayed the diffusion of steam-heated boilers for silk reeling, even though they offered immediate gains in productivity and quality. The Gensoul method of steam heating was first introduced in Lombardy in 1815, but owing to the high cost of fuel as late as 1855 only 144 of Lombardy's 3,088 silk-winding mills used steam-heated boilers.[18]

In the case of the cotton industry, high fuel costs were aggravated by the need to import raw cotton and machinery. The additional costs were partly offset by the falling cost of British yarn, and as in other parts of Europe imported yarn provided new opportunities for the expansion of mechanized spinning and weaving (although this remained largely artisan). Despite these obstacles, there were twenty-six mechanized cotton spinning mills operating in Lombardy by the early 1840s, one of the largest being Andrea Ponti's water-powered mill in Milan. There were also attempts to adopt mechanized processes in other textile industries: Cremona, which lay at the heart of one of the most prolific flax producing regions in Italy (between the Mincio and

---

[17] F. Ramella, *Terra e telai; sistemi de parentela e manifattura nel Biellese dell'Ottocento* (Turin, 1984).
[18] Greenfield, *Economics and Liberalism in the Risorgimento*, p. 111.

the Adda on the left bank of the Po) had two mechanized linen spinneries by the 1840s.

One of the most important technological advances came in the woollen industry, which like cotton was mainly reliant on imports of raw materials because of the poor quality of local wools. In contrast to the artisan weaving industries of the Biella region, in the hill districts of the Veneto to the north of Vincenza and Treviso factory-based wool production made significant progress in the first half of the nineteenth century. Unlike the Biellese, this area witnessed a rapid decline in the number of domestic producers and the rise of two large centralized enterprises: the Marzotto company in Valdagno and the Rossi factory at Schio. Both Marzotto and Rossi began as out-putters, but started investing heavily in mechanized spinning machinery after the Napoleonic wars. In 1846 Alessandra Rossi introduced mule-jennies into his factory at Schio, and thereafter all spinning was done in the workshops. To overcome the limitations of water power he switched to steam power in 1848 and his factory quickly became one of the first vertically integrated industrial operations in Italy, specializing in the production of high-quality worked and combed woollen fabrics.

The Rossi enterprise was to be one of the great success stories of Italian entrepreneurship in the nineteenth century. Although Rossi had certainly been mindful that power-driven mules would enable him to use the cheap female labour that was abundant in the rural districts around Schio, his investment was motivated primarily by technical and commercial considerations, in particular the need to establish a footing in the highly competitive markets for specialized and high-quality cloths. None the less, the high investment in technology did not pay a quick or easy dividend. Because there was no local machine-building industry, the maintenance and repair of the machinery posed major problems. The mechanized looms also required stronger wool than was available locally and thereby increased the need to purchase more costly raw materials.[19]

Where labour costs were low, the comparative advantages of costly technologies were few and the incentives slight. This was well illustrated in the case of another Venetian wool manufactory sited at Follina, which had been fitted out extensively with English, German, French and Belgian machinery. According to its owner this had 'displaced 1,666 persons' and so reduced his production costs: 'taking into account those who are saved on the looms which now for the most

---

[19] Ciriacono, 'L'industria a domicilio nel Veneto', pp. 583–6.

part occupy one operative instead of two: the improvement of the fulling mills and the presses for smoothing the cloth, which save labour by speeding up work'. But all the new equipment was hand-driven, and while fulling, pressing and shearing were carried out in Signor Colle's works, most of the spinning and all of the weaving was still done on an out-putting basis.[20]

Even in regions like Piedmont and Lombardy where both manufactures and infra-structural conditions – most notably communications – were more developed than in any other part of Italy before unification, technological progress remained patchy and uncertain. High costs of raw materials other than those available locally, the absence of indigenous supplies of combustible fuels and the abundance of cheap labour were all instrumental. But these were not the only factors. Government policies also played their part, and in the case of Lombardy and the Veneto import duties on machinery designed to protect Austrian industries increased investment costs and made competition with advanced manufacturing regions within the Austrian Empire, especially Bohemia, very disadvantageous. No less important was the uncertainty and unreliability of markets, while the general increase in production tended to push the cost of raw materials upwards – something that was especially important for industries like cotton and wool that relied on imports for their raw materials. These uncertainties of the market were also reflected in the widespread use of exclusive patents for importers of new machinery, which tended to inhibit the transmission of technologies.

This in turn reflected the weakness of internal demand and explains why some of the most 'modern' industrial ventures of this period were often closely tied to, and dependent on, public rather than private commissions. This was most obviously the case for the first engineering works. The earliest of these were closely dependent on navy commissions, and the two largest engineering works in Italy at the time of unification were the naval dockyards at Sanpierdarena outside Genoa and Pietrarsa at Naples.

The building of the first railways saw some expansion in the sector and by 1855 there were seventeen engineering shops in Milan. These were producing a range of products: small hydraulic motors and spares, machines for processing sugar and tobacco, spinning and winding mills for the silk industry, boilers, oil and flour presses, threshers, railway equipment and looms. On the other hand, they

---

[20] Greenfield, *Economics and Liberalism in the Risorgimento*, p. 105.

could not meet demand for larger machines or produce centrifugal pumps, self-acting machines, spinning jennies, railway locomotives, rolling stock or rails.[21]

Although new technologies had been adopted in a range of different manufacturing operations on the Italian peninsula before 1860, the Italian states remained essentially importers of technology and no progress had been made either in machine-building or in other industries capable of reproducing or creating applied technologies.

Technology and industrial growth after unification

The first decades: 1860–1880

The engineering industry illustrated particularly vividly the specific combination of constraints that continued to impede technological development in Italy even after unification. A fundamental handicap was the lack of adequate supplies of suitable mineral ores, so that the engineering enterprises had generally to import pig-iron from abroad. Even for those that could rely on state orders, commissions were uncertain and irregular. This in turn made specialization almost impossible, and survival depended on the ability to produce a range of different products. The weakness of internal markets again played its part here, and despite the continuing importance of agriculture it is noticeable that neither agriculture nor the food-processing industries created any significant demand for equipment that required new technologies – at least not until very much later.

Lack of specialization was reflected in the absence of a machine-building industry, which meant that manufacturers who installed new machinery had either to build their own repair shops or to rely on foreign suppliers and technicians. The limited product-range of those engineering shops that did exist further exacerbated the situation and in turn meant that the workshops tended to rely on existing artisan skills – if these were insufficient, they imported skilled labour from abroad. As Alessandro Rossi pointed out to his critics, the Italian entrepreneur had little choice but to buy abroad since foreign machines were not only better and more reliable but also cheaper than those made in Italy.[22]

---

[21] Ibid., p. 115; L. De Rosa, *L'industria metalmeccanica* (Naples, 1969).

[22] R. Maiocchi, 'Il ruolo della scienza nello svillupo industriale italiano', in G. Micheli (ed.), *Storia d'Italia*, Annali 3 (Turin, 1980), p. 887.

Many industrialists believed that the shortage of skilled labour with relevant technical training was a major factor in Italy's continuing technological backwardness, and in the decade after unification this became a subject of major debate. In 1841 and 1843 the Austrian government had opened technical schools in Milan and Venice, but had provided little support for these. In 1851 a group of Milanese manufacturers founded the Society for the Encouragement of Arts and Crafts in Milan, which offered specialist training courses and led on to the creation of the Advanced Technical Institute, which in turn became the Milan Polytechnic. In Turin private enterprise also lay behind the School of Engineering, which was founded in 1862. But most engineers received their training through university courses in physics or mathematics.[23]

The Casati Law of 1859, which unified and reorganized the Italian education system, set up Schools of Applied Studies (*Scuole di Applicazione*), which unlike the French *grandes écoles* and the German *technische Hochschulen* were part of the university system. The emphasis tended to be strongly theoretical and the expansion of laboratory-based experimental teaching was slow in quantitative terms and patchy in terms of content – at Milan there were courses in civil and structural engineering, but nothing on industrial chemistry before the mid-1880s. But there were important exceptions, notably in electrical engineering where Italy established a precocious footing when both the Milan and Turin Schools of Applied Studies set up specialist courses in electrical engineering in the late 1880s: between 1893 and 1904, 1,121 students completed the course at the Turin Advanced School of Electro-technology.[24]

None the less, Italian universities were widely criticized for failing to take a serious interest in applied science, and it was frequently claimed that graduate engineers were normally qualified only in general physics, and that although there was no shortage of engineers in Italy, a qualification in engineering had little practical application for any important branch of industry. But this was not entirely justified since the problem of skilled manpower was never simply one of supply and the cultural discrimination against applied science was not the only reason for the slow development of technical training. The relatively

---

[23] A. Guagnini, 'Higher education and the engineering profession in Italy: the *Scuole* of Milan and Turin 1859–1914', *Minerva*, 26 (1988), pp. 512–48; S. Soldani, 'L'istruzione tecnica dell'Italia liberale', *Studi Storici*, 1 (1981), pp. 79–117.

[24] A. Guagnini, 'The formation of Italian electrical engineers: the teaching laboratories of the Politecnici of Turin and Milan 1887–1914', in F. Cardot (ed.), *1880–1980: une Siècle d'Electricité dans le Monde* (Paris, 1987), pp. 283–99.

slow pace of technological development meant in turn that demand was weak, which, as one industrialist who was keenly committed to improving the skills of the Italian labour force noted, created a vicious circle that was very difficult to break: 'Technical training cannot be made to expand by the government unless the economy is also progressing. You cannot create a skilled labour force for the textile industry in a country where the textile industry does not have factories that could employ people with such skills.'[25]

The low demand for technology was also reflected in the fact that the major inventions that did take place in Italy in the decades after unification were only exploited abroad, the best examples being Pacinotti's work on electro-magnetism which was literally stolen by a French competitor; the Barsanti and Matteucci steam engine; and Guglielmo Marconi's long-distance message transmissions. Marconi's case was the most spectacular, as his inventions were turned down by the Italian Ministry of Posts and Telegraphs in 1896 as being of no interest.[26]

In the 1880s, the pace of industrial expansion in Italy began to quicken markedly and the technological shortcomings of Italian industry became increasingly visible. Expansion was fastest in the textile sectors, but even the rapidly growing cotton sector continued to depend on foreign machinery and created few demands for further technological development.

The situation was different for the chemical industry, for engineering and for metallurgy, however. But even in these cases, attentive observers were alarmed by the very slow development of new technologies. Despite the rapid advances that were being made in the application of scientific research to the development of the chemical industry in Germany from the 1860s onwards, for example, its Italian counterpart remained uncompetitive and geared to producing primary materials, locked into a self-repeating circuit of low technology, poor quality and low demand.

The development of the railway network in the decade after 1860 offered major opportunities for the development of the engineering and metal-working industries, but it is generally agreed that these opportunities were in large part wasted. Initially the demand was mainly for construction, for rails, points, maintenance and repairs – demands which did not require advanced technology. But locomotives were another matter and the development of an Italian locomotive-

---

[25] F. Carli (quoted), in Maiocchi (1980), p. 882.
[26] Ibid., p. 910.

building industry proved far from easy. Between 1861 and 1884, 1,296 locomotives were purchased by the Italian railway companies. Of these only 231 were built in Italy, mainly in the later part of the period, representing 18 per cent of the total; these came principally from two suppliers: 29.5 per cent from the Ansaldo works at Sanpierdarena and 64 per cent from the Neapolitan Pietrarsa works.[27]

The French historian Michel Merger has shown how the Navy Ministry's enquiry into the engineering industry in 1881 revealed that all the major Italian engineering construction enterprises were unprofitable, financially precarious and uncompetitive. They lacked specialization, adequate levels of technology and any form of applied research. The shortcomings of the national iron and steel industry were such that not only did the primary materials for railway construction (coal, mineral ores, copper and brass) have to be imported, but so too did semi-finished products and components like pig-iron, steel laminates and plates as well as axles, wheels, wheel rims and springs, since they could not be produced competitively in Italy.[28]

Nor were these problems dispelled by the establishment of a national steel industry when the government provided support for the building of the first steel mills in Italy using integrated production processes at Terni in 1884. The Terni works depended on government commissions and benefited from tariff protection, which has been taken to explain the decision to adopt the Martin-Siemens furnaces despite their technological deficiencies. In contrast to more advanced processes, the Martin-Siemens furnaces were cheap to operate, required less skill on the part of the workers, and rather than high quality ores could use old railway lines, which were the cheapest form in which Italy could obtain iron. As a consequence the foundry could not produce a range of alloys and specialized steels, even though the government had to place its order for armour plating for the Navy at Terni, despite the fact that its own tests demonstrated that this was inferior to Krupp plating and indeed almost completely ineffective.[29]

The period after 1880

Even twenty years after unification, and despite the new signs of

[27] M. Merger, 'Un modello di sostituzione: la locomotiva italiana del 1850 al 1914', *Rivista di Storia Economica*, new series, 3 (1986), pp. 72–3; S. Fenoaltea, 'Italy', in P. O'Brien (ed.), *Railways and the Economic Development of Western Europe 1830–1914* (London, 1983).
[28] Merger, 'Un modello di sostituzione', pp. 78–9.
[29] Maiocchi, 'Il ruolo della scienza', p. 898.

vitality that were evident in the Italian economy, it still seemed to be dogged by technological backwardness. The development of its advanced sectors, like chemicals, engineering and iron and steel, continued to be characterized by low levels of technology, a lack of contact between scientific research and industrial production, and a lack of specialization. But the twenty years that followed saw important developments which without entirely reversing the situation did bring about a radical transformation of a number of key sectors. The reasons why and how this happened also throw light on the factors that had previously encouraged technological backwardness.

One important success story, as Michel Merger has shown, was in railway engineering. Despite slow and uncertain growth until the end of the 1870s, matters soon began to change. By 1905, when the Italian railways were nationalized, the level of mechanical engineering had progressed by leaps and bounds and the workshops of the new Italian Railways Company were producing locomotive prototypes that were as advanced as any in Europe.

How had this come about? The development of specialized research and development units within the railway companies played an important role but, as Merger shows, this was more a consequence of deeper changes affecting the environment in which the major mechanical engineering and construction firms were operating by the end of the century. The first major breakthrough came with the railway law of 1885, which gave preference to national products. This was critically important because although engineering – unlike textiles, steel and agriculture – was not included in the high protective tariff regime that was introduced in 1889, the combination of renewed demand for further extension of the railway network and the government's preferential policies on Italian-built locomotives and rolling stock reduced the market uncertainties that had previously hampered locomotive construction.

The new expansion took much more specialized lines. The old Pietrarsa works ceased making locomotives altogether to specialize in general repairs, for which they were much better suited. The Ansaldo company, under the dynamic leadership of Ferdinando Maria Perrone, began to invest heavily in steel-making and modernized its engineering shops. A number of new specialist engineering works were also established – most notably Franco Tosi & Co. (1881) and Ernesto Breda & Co.. (1886), which specialized in 'mechanical construction and especially building steam locomotives, boilers, railway and tramway equipment'. Breda took over the premises of the former Elvetica

Co. and invested heavily in new machinery, most of which was imported from the USA.[30]

The preferential policy for Italian-produced goods also led a number of foreign, and especially German, companies to set up workshops in Italy. One example was the Keesler company of Esslingen which financed the Saronno mechanical engineering firm in 1887. Equally important, Merger argues, was the active participation of the new mixed banks in this sector after the banking crisis of the late 1880s and early 1890s had been overcome. The Banca Commerciale had held a major financial stake in the Breda company since much earlier, but in 1899 became the principal shareholder and financed a new spurt of expansion. The Credito Italiano, the other leading mixed bank of the 1890s, also invested heavily in the sector and was the major shareholder in the S.A. Officine company (formerly Miani, Silvestri & Co.), enabling the conglomerate to absorb a number of smaller companies. Ansaldo was one of the few major engineering construction firms to remain under family control, but in 1901 the Perrone and Bombrini families established close links with the W.G. Armstrong and Whitworth company.[31]

By 1900 Italy was self-sufficient in locomotive production and the example serves to demonstrate that once conditions had been established in which there was both regularity of demand and access to sufficient capital, technological innovation ceased to be a major obstacle. The expansion of the mechanical engineering industries in turn created a new demand for training and skills which was reflected in a rapid increase in technical schools and evening classes in the cities where the engineering industries were located – although the initiative for these was still overwhelmingly private. None the less, it was not until 1895 that the first laboratory-based courses in mechanical engineering were offered at the Milan School of Applied Studies.

The development of the Pirelli rubber concern followed a similar pattern. The company was founded in the 1870s by Giovanni Battista Pirelli, but it was not to expand significantly until the growth of the electro-technical industry created massive new markets for rubber-coated cables – a field in which Pirelli quickly established an export market and also set up subsidiary production companies abroad to become 'Italy's first multi-national'.[32]

[30] Merger, 'Un modello di sostituzione', pp. 80–95.
[31] Ibid., pp. 86–94; P. Hertner, *Il capitale tedesco in Italia dall'unità alla prima guerra mondiale* (Bologna, 1984), pp. 33–4.
[32] Toniolo, *Economic History of Liberal Italy*, p. 108.

The great success story was the electro-technical industry. As late as 1885 this had been virtually non-existent, but the absence of coal deposits on the peninsula made Italian entrepreneurs sensitive to the opportunities offered by electrical energy. The first electrical engineering plant (the Centrale di Santa Radegonda) was built in Milan in 1883 with backing from some of the city's leading industrial interests – Giuseppe Colombo, and the Crespi (cotton) and Richard (ceramics) concerns. The same group was among the leading investors in the first Italian electrical company, the Società Edison Italiana.

Electricity generating stations were built in quick succession in many Italian cities and there was a rapid expansion in hydro-electric generating plants. But the development of an Italian electrical engineering industry proved to be much slower and until 1914 Italy continued to import most of the machinery (alternators, dynamos, transformers and electrical engines) used in the production of electrical energy. Contemporaries blamed the high tariff regime for this, pointing out that the engineering industry enjoyed no protection while import tariffs kept steel prices high without leading to quality improvements. The growth of all branches of the engineering industry was as a result handicapped by high material costs and poor quality inputs.[33]

Since there was little change after the tariff regime was modified in 1903, historians have argued that these charges were exaggerated. But both contemporaries and historians have been aware that in the years after 1890 foreign capital also played a part in retarding technological development in certain sectors, since its main concern was to increase markets for its own products. It was widely recognized that the German consortia that were the principal investors in the Italian mixed banks – AEG, Siemens and Schuckert (after 1903 Siemens-Schuckert) – were interested in expanding the Italian electro-technical industries since they provided markets for German electrical engineering products: 'German capital investment in Italian hydroelectric plant is not an end in itself but rather a means to sell more German machinery.'[34]

In other sectors, the complex structure of company finances also

---

[33] Maiocchi, 'Il ruolo della scienza', p. 896; Guagnini, 'Formation of Italian electrical engineers', pp. 284–5; M. Warglien, 'Nota sul'investimento industriale in maccinari ed attrezzature meccaniche: Italia 1881–1913', *Rivista di Storia Economica*, new series, 2 (1985), pp. 125–139; Toniolo, *Economic History of Liberal Italy*, pp. 69 and 108.

[34] Quoted in Hertner, *Il capitale redesco*, p. 145; also R.A. Webster, *Industrial Imperialism in Italy 1908–1915* (Berkeley, 1975); V. Zamagni, *Industrializzazione e squilibri regionali* (Bologna, 1978); J.S. Cohen, 'Italy 1861–1914', in R. Cameron (ed.), *Banking and Economic Development. Some Lessons of History* (London, 1972).

served to impede technological development and especially investment in independent research and development. Even after the turn of the century few Italian universities offered courses in the applied sciences, and the largest steel producers in Italy were still content to import German, French or British equipment, know-how and technical advisers. In new industries like motor car construction firms like FIAT were imitating American state-of-the-art production techniques, but research and development continued to be an *ad hoc* and largely artisan affair with little systematic application of science to production.[35]

Conclusions

The overall balance-sheet of technological development and innovation in Italy by the eve of the First World War was still uneven. In some areas Italian industrialists had succeeded in moving to the forefront of technological development, demonstrating that there were no absolute barriers. But these cases were few, and on balance Italy continued to depend on foreign producers to meet its needs for technology. This was true of traditional industries like cotton-spinning as well as the new electro-technical industries. The continuing preference for foreign technology necessarily stunted the development of a broad machine-building sector and room for expansion was further limited since neither agriculture nor civil construction created a significant alternative demand for advanced technologies. Hence the demand for technical skills in most sectors of Italian industry remained narrow and weak.

There were exceptions, as the electro-technical industries demonstrated. Demand was not totally lacking, as the steady but still very slow and partial expansion in technical training shows. But relatively few Italian industries were employing trained scientists on the eve of the Great War, and the universities had done little to change their rooted aversion to the applied and bench sciences. Since demand for technical skills was weak, the shortcomings of higher education cannot be held entirely to blame. As we have seen, the presence and preferences of foreign capital by the turn of the century exercised important constraints on the transmission of innovation from one sector to another.

No less important were the internal constraints posed by the persistence of a low wage economy in which cheap labour continued to be one of the most important comparative advantages for Italian

---

[35] Maiocchi, 'Il ruolo della scienza', pp. 918–19.

producers – indeed in many cases their only comparative advantage. Despite emigration, labour was abundant and Italian employers repeatedly resisted moves that might change this situation. Technologies that displaced skilled labour in favour of unskilled were therefore welcomed, whereas those that created a demand for more technically prepared – and hence more costly – labour were avoided.

The prevalence of a low wage economy was in itself a further constraint on the broadening of internal markets, and especially internal markets for high-technology products. This did not constitute an absolute barrier to technological progress, but it did make broader processes of innovation difficult and helps to explain why the technical structures of Italian industry on the eve of the First World War shared in unequal proportion the features of both the first (low-technology) and the second (high-technology) revolutions.[36] Although it had proved possible in certain sectors to create conditions that permitted rapid and successful technological development, the overall structure of the economy did not facilitate the transmission of technological advances from one sector to another to stimulate deeper processes of innovation and economic growth. None the less, by importing foreign technology the Italian economy had succeeded in taking gigantic steps forwards after 1880. In terms of productivity there was a price to pay for this dependence, of course, but possibly the most serious consequences of the persistence of low levels of technology in industry and in the labour force lay in the constraints this imposed on the growth of internal markets – weaknesses that were not only economic.

---

[36] Cafagna, 'Industrial revolution in Italy 1830–1914', pp. 322–4.

6
Technical and Structural Factors in British Industrial Decline 1870 to the Present

Derek H. Aldcroft

Innovation and growth

There is fairly general agreement that innovative activity and technical progress in the widest sense of the term are of crucial importance to high-income industrial countries. After all, innovative activity is, as Kerry Schott points out, virtually the only comparative advantage left to the developed world.[1] Some economists would even argue that technology, or technical change, is the chief, if not the only, determinant of economic growth. Usher, for instance, concludes that economic growth depends predominantly on technical change and that it cannot occur to any significant extent in its absence.[2] While this may be something of an extreme view, and is one that has been criticized on a number of occasions, the fact remains that countries which fail to keep pace with the rate of innovative activity set by the leaders in best practice techniques will lose market share (both at home and abroad) and in the long run they will degenerate into low-wage service economies catering mainly for the domestic market.

For a country that persistently fails to product-innovate there is no obvious escape route, as Stoneman has reasoned. Devaluation and

---

[1] K. Schott, *Industrial Innovation in the United Kingdom, Canada and the United States*, British-North American Committee (London, 1981), p. 65.
[2] D. Usher, *The Measurement of Economic Growth* (Oxford, 1980), p. 289.

wage cuts may ease the situation for a time but they are not likely to rejuvenate a traded-goods sector which is continually under threat from superior foreign product innovation. What is more likely to happen is that resources will tend to shift into the sheltered non-tradeable activities and eventually the country will become a low-wage non-trading economy. Stoneman believes that Britain's poor technological record has been a major factor in her economic decline though he admits that its contribution is difficult to assess because of intermittent compensations through wage and exchange rate changes.[3]

In several respects this scenario provides a reasonably good fit for the British economy during the past hundred years. Apart from short intermissions (e.g. the 1930s) the relative economic position of Britain has been slipping since before the turn of the century. Once the second highest income country in the world (after Australia), it had by the 1970s become one of the poorest of the advanced industrial nations according to international rankings based on domestic output per capita. The speed of the decline was particularly marked in the super-growth era of the postwar years when Britain's rate of output growth was less than two-thirds that of her advanced competitors.[4] International trade data for manufactured goods demonstrate Britain's weakening position even more sharply. From a position of strength, accounting for some one-quarter of the value of world exports in manufactured goods, Britain's share steadily dwindled, to less than 10

---

[3] P. Stoneman, 'Technological change and economic performance', in *Out of Work: Perspectives of Mass Unemployment* (Department of Economics, University of Warwick, 1984), pp. 56, 61. A good micro illustration of the futility of wage cuts as a solution to declining competitiveness because of technical backwardness is provided by Courtauld's Maple 2 spinning mill in Oldham, near Manchester, which is currently being modernized with new and more efficient machines at a cost of £4.5 million. Labour productivity at the mill is expected to double and employment will fall by more than half, from 250 to 100 workers, but the remaining employees will benefit from increased pay. The new technology is dictated by the need to keep up with competitors and to supply a better product to the customer; otherwise the mill would go out of business since its present equipment is outdated and uncompetitive. The question of wage cuts scarcely enters the picture since there is virtually no level of wages which could provide a substitute for investing in new and more capital-intensive technology to supply the needs of the market and thereby remain internationally competitive. See M. Prowse, 'When paying workers less doesn't help', *Financial Times*, 13 March 1986. Watchmaking also provides an excellent example of how the failure to product-innovate – from mechanical to electronic timepieces – can render an industry obsolete virtually overnight, as Timex in Scotland discovered to their cost. The Swiss have also lost out in the latest round of technology in this field.

[4] For comparative data see D.H. Aldcroft, 'Britain's economic decline 1870–1980', in G. Roderick and M. Stephens (eds), *The British Malaise* (Lewes, Sussex, 1982), pp. 32–5.

per cent in the 1970s. Figures for import penetration tell a similar story. Imports as a percentage of home demand doubled between 1955 and 1969 (from 8 to 16 per cent) and then almost doubled again to reach 30 per cent by 1980.[5] Not very long afterwards the balance of trade in manufactured goods swung into deficit for the first time in recorded history, a shift of no little concern to the recent House of Lords Committee on Overseas Trade.[6] Other indications that Britain fits the pattern described above are the trend away from manufacturing and towards services and the fact that this country has become a low-wage economy compared with her main competitors.

Rosenberg has cautioned against laying too much stress on trade share data for the purpose of demonstrating Britain's deteriorating performance, since he argues that declining relative importance in world trade shares in manufactured goods was only to be expected given rapid industrialization abroad and supply constraints at home. What he sees as of greater significance is Britain's seeming inability

> to generate or exploit new technologies with anything like the success achieved in coal, iron, and steam technologies of the industrial revolution ... in the twentieth century, the British economy responded slowly and sluggishly to the structural transformation required by the decline of the old industries and the expansion of the new ones. Industries such as textiles, coal, steel, and shipbuilding failed to achieve either sufficient reduction in size or technical modernization, where the latter was possible. The pressure to make such structural transformations became increasingly urgent in the twentieth century as wars, depression, and the growth of import substitution abroad drastically changed the export markets for many of the traditional older industries.[7]

In some respects Rosenberg's thesis rather avoids the question since Britain's poor performance in international trade partly reflects the internal weaknesses of the economy which he outlines. Moreover, not all the decline in trade share can be accounted for by the industrialization of new-comers to the economic growth scene since some of the early-starters, for example Germany, have done very much better in hanging on to their shares in the postwar period. However, he does recognize that there is more to the British decline than mere

---

[5] Data from K. Williams, J. Williams and D. Thomas, *Why Are the British Bad at Manufacturing?* (London, 1983), pp. 114, 118.

[6] *Report from the House of Lords Select Committee on Overseas Trade* (London, 1985), H.L. 238-I.

[7] N. Rosenberg, *Inside the Black Box: Technology and Economics* (Cambridge, 1982), pp. 258–9.

technological backwardness. There is, as he stresses, an important structural dimension which has to be taken into account and which assumes particular importance in the period through to the first half of the twentieth century. If we recast the British problem of the last century it would look something like the following: (a) a structural handicap through to the 1930s, though diminishing over time; (b) the fossilization of old sectors, which continue to act as a drag on the economy and to which may be added an increasing market structure problem; and (c) an across-the-board deficiency in industry post-1945.

Structural handicap

The structural factor is particularly pronounced in the period before the First World War. Before 1914 there were relatively few major structural changes in the pattern of economic activity. By and large, the continued expansion of the economy depended upon the development of existing sectors. This was especially true in the case of the tradeable (industrial) sector, which accounted for around 45 per cent of total domestic ouput. A few large staple industries dominated the field. In 1907, coal, textiles, iron and steel and engineering including shipbuilding accounted for 50 per cent of net industrial output, employed about one-quarter of the occupied labour force and supplied no less than 70 per cent of Britain's exports.[8] Resources continued to be poured into these sectors despite the fact that their rate of expansion (especially productivity) slackened in the couple of decades or so before the war. Low productivity growth in these industries was primarily a function of a slow rate of technical advance, while developments in new high-growth sectors based on inventions such as electricity and the internal combustion engine were very limited. New industries (electrical engineering, road vehicles, rayon, chemicals and scientific instruments) accounted for only 6.5 per cent of net industrial output and 5.2 per cent of industrial employment.[9] Thus as far as industry was concerned the shift of resources from low- to high-growth sectors was very limited. The service sectors of the economy – transport, finance, distribution etc. – continued to expand, especially in terms of employment, where they increased their relative importance. But the rates of output and productivity growth of these sectors were relatively low and therefore could not fully offset the retardation in the industrial

[8] On A.E. Khan's definition in *Great Britain in the World Economy* (New York, 1946), pp. 66–7.
[9] Ibid., pp. 106, 109.

sector. It should also be noted that even at this stage Britain was losing trade share in most sectors, whether expanding, stable or declining in terms of international trade, so that her problems were not confined solely to the old staple trades.[10]

The main issue here revolves around the question of what scope there was for improving economic performance through resource reallocation. Some years ago McCloskey argued that the economy was doing as well as could be expected given its resource endowments and that there was little scope for improvement through structural readjustment, including a shift of resources from foreign to home investment.[11] Unfortunately his analysis lacked dynamic properties and it assumed unjustifiably that factor supplies (capital and labour) were inelastic. Removing these constraints makes the world of difference to the potential for growth, as Kennedy has shown. His preliminary calculations suggested that had Britain made a commitment of resources to telecommunications, electricity, engineering (including electrical), car manufacturing, construction and related industries similar in extent to that in the United States, then the implied increase in growth would have been sufficient to raise British per capita incomes to 55 per cent above the level actually recorded in 1913. In a subsequently more detailed study of structural change and resource reallocation for this period, Kennedy set out alternative growth profiles contingent on a greater commitment of resources to those sectors capable of sustaining rapid expansion and technological progress. The upper bound growth potential is substantial, producing up to a doubling of the realized per capita income of 1913. It should be noted that the structural change envisaged to generate such income growth is very considerable, involving a commitment of resources to dynamic sectors greater, in relative terms, than in the United States during the same period. But such a projected reallocation was by no means outside the bounds of possibility given earlier historical experience during the industrial revolution.[12] It would also have involved a

[10] H. Tyszynski, 'World trade in manufactured commodities, 1899–1950', *The Manchester School*, 19 (1951), p. 283.

[11] D.N. McCloskey, 'Did Victorian Britain fail?', *Economic History Review*, 23 (1970); see also the critique by N.F.R. Crafts, 'Victorian Britain did fail', *Economic History Review*, 32 (1979).

[12] W.P. Kennedy, 'Institutional response to economic growth: capital markets in Britain to 1914', in L. Hannah (ed.), *Management Strategy and Business Development* (London, 1976), p. 183; 'Economic growth and structural change in the UK, 1870–1914', Discussion Paper 112 (1978), Department of Economics, University of Essex; 'Foreign investment, trade and growth in the United Kingdom, 1870–1913', *Explorations in Economic History*, 11 (1974).

significant switch of resources from foreign to domestic investment,[13] a shift which, it is argued, would have been economically rational given that most foreign investments yielded less than the average domestic return.[14]

The fact that Victorian and Edwardian Britain failed to shift to a more dynamic economic structure is less easy to explain, despite the numerous writings on the subject of retardation. Kennedy himself reasons in terms of the imperfections of the British capital market, which tended to concentrate on fixed interest issues and avoided new and more risky technological developments. In other words, the institutional financial mechanism acted as a constraint and thereby led to an ossification of the existing structure. Moreover, the fact that the large proportion of savings flowing into foreign investment restricted the growth of the domestic economy in turn helped to condition the behaviour of entrepreneurs.[15]

The structural argument does not, of course, preclude a technological lag before 1914. Indeed, several of the older industries, notably coal, steel and textiles, have been criticized for their slowness to innovate, although econometric studies have tended to excuse them on the grounds that by taking a short run profit profile they were acting rationally. Nevertheless, even in some of the newer, potential-growth sectors there is evidence of sluggish innovatory response which is less excusable. The chemical industry is a good case in point. Despite the fact that much pioneering developmental work on the Solvay process of alkali production was undertaken by Mond in Lancashire, British producers allowed the lead in production to be taken by Germany. Between 1890 and 1900 British exports of alkalis fell by about one-half, largely as a result of a failure to match the German take-up of the new process. In 1882 and 1900 the Solvay process accounted for 44 and 90 per cent respectively of German soda production as against 12 and 40 per cent in the case of the British. Partly because of the capital and skills already locked into the old Leblanc process, British manufacturers reacted defensively and responded by trying to stretch the product life cycle of the Leblanc process.[16]

Thus technological shortcomings were certainly apparent before

[13] Kennedy, 'Foreign investment'.

[14] Though see M. Edelstein, 'Realised rates of return on UK home and overseas portfolio investment in the age of high imperialism', *Explorations in Economic History*, 13 (1976).

[15] Kennedy, 'Foreign investment', p. 439.

[16] R. Rothwell and W. Zegveld, *Reindustrialisation and Technology* (London, 1985), pp. 40–1.

1914 but the main problem was a structural one: that is, too many resources concentrated in slow-growing sectors with a low rate of technical advance, while large scale innovations, particularly those giving rise to new industries, were slow to mature. Inter-sectoral shifts of resources tended to be between low productivity growth sectors rather than from low- to high-growth sectors.

The contrast with the postwar situation is quite marked. In the first place the export-predominant staple industries declined in importance and gave way to the new industries, which expanded rapidly by exploiting new techniques, and which reaped scale economies and had an above-average rate of growth. In 1935 the new industries accounted for 19 per cent of net industrial output compared with 12.5 per cent in 1924 and 6.5 per cent in 1907. These figures take no account of electricity supply, which expanded very rapidly in the interwar years. Secondly, the construction and allied trades experienced substantial growth, largely as a result of the massive housebuilding programme. Thirdly, although rail and tram transport stagnated, this was more than compensated for by the vigorous boom in motor transport. Finally, the service sectors of the economy continued to expand, especially in terms of employment.

As far as the tradeable sector is concerned, there was some relative improvement in the 1930s compared with the previous decade. British trade shares stabilized on balance and gains were made in half the sixteen major commodity groups listed by Tyszynski.[17] Although in some sectors Britain still remained competitively weak, there had on average been some improvement in competitiveness due in part to structural change and an acceleration of technical advance, especially in the newer sectors. As time went on the commodity composition of Britain's export trade improved so that by the end of the period there is little evidence that it lagged seriously behind the structural shifts in world trade.[18]

This proved to be only a short interregnum in Britain's long-term decline. It is true that the structural format of the economy improved during the interwar years and through the 1940s when war gave a boost to future potential growth industries such as chemicals, electrical engineering and instrument manufacturing. By the early 1950s Britain no longer had a serious structural handicap compared with, say, Germany. Unfortunately, new handicaps emerged which were to

---

[17] Tyszynski, 'World trade in manufactured commodities', p. 249.
[18] Kahn, *Great Britain in the World Economy*, p. 133; A. Maizels, *Industrial Growth and World Trade* (Cambridge, 1963), pp. 93, 226–31.

plague the postwar economy. First, the artificial conditions of the 1930s, when nationalistic policies were much in evidence as a response to the great depression, led to a shift in market structure for Britain's exports towards imperial and colonial markets, a trend which continued in the difficult years of the 1940s when Britain turned her back on Europe.[19] For much of the postwar period this meant a concentration on downstream markets in terms of growth and technological spin-offs, as opposed to the stimulation that would have been engendered by focusing on the rich and expanding markets of Europe and North America. Through the 1950s and 1960s only about one-third of Britain's manufactured exports on average went to EEC countries and North America.[20] The impact was reflected in the unit values of Britain's traded goods: relatively low unit values for exports and high unit values for imports. Thus, having sloughed off one structural handicap Britain managed to take on board a new one. Secondly, the sellers' market in the early postwar years, coupled with the gravitation towards the safer and protected imperial markets, engendered a complacent attitude on the part of businessmen in terms of competing in the high-technology markets. Thirdly, the protective conditions of the 1930s meant that ailing industries failed to collapse on schedule, and they were given a new lease of life in the war and its aftermath. By the end of hostilities they were antiquated and played-out and they became a perpetual millstone throughout the postwar period when governments channelled massive resources into them in a futile effort to protect their employment position. Correlli Barnett[21] has painted a gloomy picture of the traditional sectors as follows:

> The record of the Second World War thus demonstrates Britain's great traditional industries to have indeed suffered from the same kind of weaknesses that brought about the collapse of the French Army in 1940, from outdated technology and doctrine to poor leadership and to morale so low as sometimes to verge on the mutinous.... [the] wartime record ... not only demonstrates the shortcomings of British heavy industry, but also reveals that these shortcomings were so intractable that even the direct intervention of a government vested with extraordinary powers,

---

[19] I refer here to the postwar Labour Government's refusal to get into Europe in the later 1940s, which I recall Alan Milward emphasizes as a turning point for the worse in his *The Reconstruction of Western Europe, 1945–51* (London, 1984).

[20] The proportion did increase somewhat, from around one-quarter in the early 1950s to just over 40 per cent in the later 1960s. Williams, Williams and Thomas, *Why Are the British Bad at Manufacturing?*, p. 114.

[21] C. Barnett, *The Audit of War* (London, 1986), pp. 123–4.

even the psychological spur of desperate national danger, could do little to remedy them.

Such a record offered no legitimate hope that in the postwar era these industries would prove unqualified national assets, powerful sinews of prosperity, like their German or American counterparts, but instead offered the near certainty that they would continue to present grievous long-term national problems, costly and insoluble.

Although the main structural burden had been removed there were still serious intra-industry structural problems evident in many of the older industries, such as shipbuilding, coal and steel. In shipbuilding, for example, in both the interwar and postwar years, output was fragmented among a large number of small yards and firms, often dominated by independent and rival family cliques, and based on a system of craft specialization that was wholly inappropriate to the market conditions of the twentieth century. Ships tended to be custom-built, with little attempt at standardization, and the proliferation of jealously guarded craft skills gave rise to frequent demarcation disputes and inefficient manning levels. According to Lorenz and Wilkinson, the industry in its heyday became locked into an institutional structure, rigid and inflexible, from which it never really managed to escape. Management lacked the ability and the will to rationalize the structure of the industry and apply modern work methods. Few yards could therefore achieve a minimally efficient level of operation and no single firm was large or powerful enough to impose a solution on the rest of the industry. The independence of family-controlled concerns also militated against a co-operative solution.[22] Thus the industry experienced cumulative decay, and even nationalization in 1977 did not save it from a further sharp contraction in size.

Inefficiency and technical backwardness

The early postwar period therefore opened on a mixed note. Britain was relatively strong *vis-à-vis* Europe but not compared with the

---

[22] E. Lorenz and F. Wilkinson, 'The shipbuilding industry, 1880–1965', in B. Elbaum and W. Lazonick (eds), *The Decline of the British Economy* (Oxford, 1986); A. Slaven, 'Growth and stagnation in British/Scottish shipbuilding, 1913–1977', in J. Kuse and A. Slaven (eds), *Scottish and Scandinavian Shipbuilding Seminar: Development Problems in Historical Perspective*, University of Glasgow (1980). For similar problems in Scottish steel see P.L. Payne, *Colvilles and Scottish Steel Industry* (Oxford, 1979).

United States. It also had a more favourable product structure than hitherto. It faced a sellers' market for a time but its export market structure was unfavourable for the longer term. It still had a residue of old industries with internal structural problems, some of which (iron and steel, coal and shipbuilding) faced good markets for a decade or so after the war.

The advantages, such as they were, soon disappeared and taking the postwar years as a whole British manufacturing steadily lost ground. Its competitive deterioration is quite easily demonstrated by the almost continuous loss of market share in nearly all sectors of activity, which is reflected in the adverse trading elasticities for both exports and imports. It shows up too in the unit values for tradeable goods: that Britain increasingly imported higher unit value products (more technologically sophisticated) and exported relatively low unit value goods. These features are consistent with what is known about the market structure of British trade. This situation would arise because of a slow rate of innovation, which would lead to relatively low productivity growth, higher relative unit costs (assuming real wage growth similar to elsewhere and a constant exchange rate) and a less sophisticated range of products on offer for export. As far as productivity growth in manufacturing is concerned, Britain consistently underperformed relative to her main competitors. From 1950 labour productivity in manufacturing industry in France and Germany increased at an annual rate 50 per cent higher than that achieved in the UK and in absolute terms per head productivity was some 50 per cent higher in these two countries than in Britain by 1971, while the absolute differential was nearly double when comparison is made with the United States.[23]

The comparison with Germany is particularly instructive since in many respects the economies of the two countries were quite similar by the middle of the 1950s. They had roughly similar levels of output and capital per person employed and the industrial structure of the two countries – in terms of the sectoral distribution of output, factor input and the size of companies and plants – was approximately comparable. Both started out from a similar level of efficiency, while throughout the period Britain maintained a favourable commodity pattern of exports (that is in sectors expanding in world trade). Yet during the

---

[23] C. F. Pratten, *Labour, Productivity Differentials within International Companies* (Cambridge, 1976), pp. xi, 17; see also A.D. Smith, W.N. Hitchens and S.W. Davies, *International Industrial Productivity: a Comparison of Britain, America and Germany* (Cambridge, 1982), p. 5.

period 1954–72 there was not one single branch of industrial activity in which Britain performed better than Germany. Germany's productive capacity expanded much more rapidly than that of Britain and Germany achieved a much greater improvement in the efficiency with which resources were used, indicating, argues Panic, a much greater absorption of new technologies. Not surprisingly, therefore, Germany's relative competitiveness improved, allowing it to raise its share of world trade in manufactures from 19.2 to 22.4 per cent between 1955 and 1973, whereas Britain's slid catastrophically from 22.9 to 7.5 per cent.[24]

Clearly, Britain's problem has been an all-pervasive one in that there has been in the postwar period a widespread relative decline in efficiency and competitiveness right across the tradeable sector (manufacturing). Broad structural deficiencies (composition of output, exports etc.) do not enter into the picture in this period since by all accounts there was little, if any, deficiency on this score, though one could be critical of the resources that have been disbursed in propping up declining sectors of activity.[25] Market structure was, however, a negative factor, at least until the 1970s when accession to the European Community brought about a shift in trade patterns.

It is more difficult to disentangle the causal sequence for this generalized deficiency in manufacturing. It would be easy enough to lay the blame on the failure to innovate. There is certainly no lack of writings commenting upon Britain's poor record in technological innovation and technical progress, nor of authors linking innovative activity with trade performance.[26] However, what emerges from much of the work on comparative productivity is that to point to a single cause for Britain's problem is far too simplistic. There are in fact

[24] M. Panic (ed.), *The UK and West German Manufacturing Industry, 1954–72*, National Economic Development Office, Monograph 5 (1976), pp. ix–xi.
[25] Ibid., p. x; Smith, Hitchens and Davies, *International Industrial Productivity*, p. 94.
[26] See, for example, K. Pavitt (ed.), *Technical Innovation and British Economic Performance* (London, 1980), p. 4; C. Freeman, 'Technical innovation and British trade performance', in F. Blackaby (ed.), *Deindustrialisation* (London, 1979); S. Golmulka in W. Beckerman (ed.), *Slow Growth in Britain* (Oxford, 1979). For individual case studies see K. Wagner, 'Competition and productivity: a study of the metal can industry in Britain, Germany and the United States', *Journal of Industrial Economics*, 29 (1980); A. Daly and D.T. Jones, 'The machine tool industry in Britain, Germany and the United States', *National Institute Economic Review*, 92 (1980); R. Rothwell and W. Zegveld, *Technical Change and Employment* (Oxford, 1979), pp. 79–93 (for textile machinery); R. Rothwell, 'The relationship between technical change and economic performance in mechanical engineering: some evidence', in A.J. Baker (ed.), *Industrial Innovation* (1979).

numerous inter-related factors at work, both economic and behavioural, ranging in intensity across sectors, and for this reason alone 'there can be no single solution to matching the performance of manufacturing industry in other countries.'[27] It is also doubtful whether some of the purely economic explanations of Britain's industrial inefficiency can be attributed or linked directly to a lapse in technological innovation. This is certainly the case with such factors as differences in the rates of output of products, length of production runs, product mix, size of plant and capacity utilization, while many behavioural forces such as strikes, restrictive practices and union structure do not readily fall into this category.[28] On the behavioural side Pratten felt that it was difficult to avoid the conclusion that in Britain there was a different attitude to authority and work, resulting in less intensity of work than in other countries.[29] On the other hand, this does not preclude the very real possibility that all these factors can conceivably influence the rate at which new techniques are taken up, which leaves us with the perennial problem in economics, namely that of the inter-relatedness of diverse factors.

Conclusion

It is not pleasant to end on an inconclusive note. What I have tried to demonstrate is that Britain's relative industrial decline since the late nineteenth century cannot be written solely in terms of a lack of technological innovation. It has certainly been an important element but it would be misleading not to recognize that there are other factors of relevance. Before 1914 the main problem was that of a skewed industrial structure which was unfavourable to growth. Had resource reallocation occurred the growth performance would have improved. The handicap of the industrial structure diminished during the inter-war period and through the 1940s, while technical progress speeded up, but at the same time market structure deteriorated. After the Second World War the structural factor no longer acts as a drag on the economy (except market structure for exports), but instead widespread and pervasive inefficiency takes hold in the tradeable sector. Whether

---

[27] Pratten, *Labour*, p. 62; Smith, Hitchens and Davies, *International Industrial Productivity*, pp. 98–9.

[28] Pratten, *Labour*, p. 71; C.F. Pratten and A.G. Atkinson, 'The use of manpower in British manufacturing', *Department of Employment Gazette*, 84 (1976).

[29] C.F. Pratten, 'The efficiency of British industry', *Lloyds Bank Review*, 123 (1977).

this can be attributed primarily and directly to technological shortcomings is a moot point given the range of conditioning factors which have been adduced for Britain's poor industrial performance. However, what seems to emerge clearly from recent history is the great complexity of issues involved and the diversity of experience among different sectors of activity, which of course makes it that much more difficult to formulate a viable and coherent policy for Britain's industrial regeneration. Nevertheless, in the context of the current situation Rothwell and Zegveld have recently argued the case for a technological transformation which they believe cannot be achieved by traditional demand management policies alone, but which will depend on technologically orientated re-industrialization policies.[30]

[30] See Rothwell and Zegveld, *Reindustrialisation and Technology*; and R. Rothwell, 'Reindustrialisation, innovation and public policy', in P. Hall (ed.), *Technology, Innovation and Economic Policy* (Oxford, 1986).

7
Crisis and Continuity: Innovation in the British Automobile Industry 1896–1986

Richard Whipp

In spite of an extensive literature devoted to car manufacture, innovation in the automobile industry remains imperfectly understood. This is especially true of Britain. An explanatory profile of the main changes in the British car industry in the twentieth century is needed since it constitutes a sector of the UK economy. The distinctive character of innovation in this key industry will be explained via an analysis of the ways in which car companies have attempted to match market with product and the changes in their resulting production and work organization practices across time.

There are four sections. The first examines the structural changes in the industry between 1896 and 1970. The industry has been transformed from a broad base with a large number of producers making a wide range of products to a structure dominated by a small group of corporations to which entry is virtually impossible. This transformation is examined with reference to three main periods of change, from which the contrasting experience of the US and British car industries becomes apparent.

The second section is concerned with the changes that occurred in

This chapter is derived from research supported by the Economic and Social Research Council.

the 1970s and 1980s. In the decades between 1930 and 1970 strong continuities of product type, production method and work organization, or rather the nature of incremental change, stand out. The 1970s and 1980s represent a period of crisis and profound change. The 1970s saw the end of three decades of growth and the crisis has led to major upheavals in almost every sphere of the car makers' activities, as competition has intensified and production has been reorganized on a global scale.

The third and fourth sections underline the particular nature of innovation in the British auto industry. In the third it is argued that to understand the experience of the UK companies we must move away from the conventional models of innovation which have been developed from the North American experience. In Britain, car-makers have evolved their own strategies of change related to a specific set of market, management and state relations. The fourth section examines the linked innovations in product, production and work organization which have made the UK car industry distinct not only from that of the USA but also from those of other European countries.

Structural change 1896–1970

The British car industry has undergone a complete transformation in the twentieth century. The change has been from a high number of producers who enjoyed ease of entry and made a large range of products to a situation where a small and exclusive set of corporations offer a limited model which is produced on a scale that makes entry prohibitively costly. How did these commercial innovations and sectoral alterations come about? To answer the question we must consider separately three main periods of change: the interwar years, the 1950s and the twin decades of the 1960s and 1970s.

Daimler was the first company to make cars in Britain in series, followed closely by Lanchester, Rover, Triumph and Riley. Basic engineering requirements and small capital outlays made entry to the trade relatively straightforward. To 1913, 198 makes of car appeared, of which 103 were discontinued. Compared with America, where standardization and mass production were the hallmarks of Ford in the 1900s, the early development of the British industry in this respect was slow.[1]

---

[1] D.G. Rhys, *The Motor Industry. An Economic Survey* (London, 1972), pp. 8ff.; R.J. Overy, *William Morris: Viscount Nuffield* (London, 1976), pp. 7, 26; S.B. Saul, 'The motor industry in Britain to 1914', *Business History*, 5 (1962).

It has been suggested that the absence of standardization and mass production in Britain can be explained by differing engineering traditions. The origins of British car production in hand-made models, as opposed to the American use of mass production almost from the outset, are also relevant. British workers had nurtured a broad craft tradition whereas in the US highly skilled workers were employed only on the more difficult jobs. Recently Foreman-Peck has indicated that by 1918 American factory design and techniques were known and available to the British car-producers but the critical difference was in reduced per capita income compared to America and hence in the reduced size of demand in Britain.[2] Moreover, during the Great War American car-makers were able to expand continuously into world markets whereas British and European production was severely disrupted.

The disruptive experiences of the two World Wars form logical boundaries to the first of our three major phases of development. Overall the number of firms was reduced from ninety-six car-producers to forty-one by 1929. The main reason for the mortality rate was the inability of the small firms to compete with the low prices of the rapidly emerging trade leaders, Morris and Austin. These two companies had secured the benefits of flowline production, albeit in its early form, enabling increased volumes and smaller, cheaper cars which began to tap a growing consumer demand. In 1929 these two companies took 88.3 per cent of total car production.[3] Within ten years, however, this position had been overturned and six major producers dominated the industry. Owing to management and organization problems at Austin and Morris, their competitors (Rootes and Standard, followed by the American firms of Ford and General Motors (Vauxhall)), were also able to meet the increasing demand for smaller cars. Instead of there being an early concentration of ownership and production in two or three organizations, as had occurred in the USA, the intervention of the American companies led to the growth of the so-called 'Big Six': Morris, Austin, Rootes, Standard, Vauxhall and Ford.'[4]

[2] H. Kerr-Thomas, as quoted in F.G. Woollard, 'Some notes on British methods of continuous production', *Proceedings of the Institution of Automobile Engineers*, 19 (1924–5), p. 467. J. Foreman-Peck, 'The effect of market failure on the British motor industry before 1939', *Explorations in Economic History*, 18 (1981), pp. 266–71.

[3] Rhys, *Motor Industry*, p. 11. A. Silbertson and G. Maxey, *The Motor Industry* (London, 1959), p. 107.

[4] Rhys, *Motor Industry*, p. 13. Foreman-Peck, 'The challenge of the twenties: multi-nationals and the European motor industry', *Journal of Economic History*, 42 (1982), p. 4.

In the postwar period, and especially during the 1950s, the dominant trends of the 1930s were maintained and even intensified. But the Second World War, like the First, was a benchmark for the industry. Only twenty firms emerged after the halt in production between 1940 and 1945. In the postwar period the domination of the Big Six was maintained. Between 1947 and 1954 they took a share of production which rose from 90 to 96 per cent. At the margin, firm life-expectancy remained short. Lotus, formed in 1953, was the only permanent and successful entrant and by the 1950s barriers to entry were considerable. Obstacles derived from size: entrants had to begin on a large scale or suffer heavy cost penalties. Existing product differentiation and consumer loyalty militated against novelty and the major manufacturers appeared to enjoy absolute cost advantages in retailing their product through prime-site networks. The McKenna duties largely prevented foreign entry. Changes within the dominant group centred on the consolidation of the Big Six to the Five following the merger of Austin and Nuffield (formerly Morris) at the third attempt since 1924.[5] This process of concentration was completed by what was regarded as the long overdue formation of British Motor Holdings in 1952. BMH, made up of Austin and Nuffield, merged with the Leyland Group in 1968; Leyland Motors, a commercial vehicle manufacturer, had purchased Standard in 1961 and merged with Rover in 1967. Although occurring in the 1960s, the formation of British Leyland (BLMC) was in essence the outcome of a series of attempts at merger which had been mounted by the same participants during the 1950s. The accompanying changes in market share are shown in table 1.

The general process of concentration arose through a mixture of forces, some highly rational and others more subjective. Personal motivations often influenced the decision-making of leading motor entrepreneurs, the best examples being the tenacity with which William Morris tried to retain personal control of his company throughout the merger with Austin and the well-documented consequences of long-standing individual and group rivalries which delayed the formation of BLMC. Some economists argue that such factors thwarted the 'free rationalization' of the industry in the mid-twentieth century.[6] In more rational terms the dominant incentive for concentration was the search for economies of scale arising from the product and component standardization which followed

---

[5] Rhys, *Motor Industry*, p. 20.

[6] J. Foreman-Peck, 'Exit, voice and loyalty as response to decline: the Rover Company in the inter-war years', *Business History*, 23 (1981), pp. 203–4.

Table 1 UK market shares, 1947, 1954, 1967 and 1973

1947	%	1954	%	1967	%	1973	%
Morris	20.9	BMH	38.0	BMC	34.8	British Leyland	32.0
Austin	19.2	Ford	27.0	Jaguar	1.4	Ford	23.0
Ford	15.4	Standard	11.0	Leyland	7.9	Peugeot	
Standard	13.2	Rootes	11.0	Rover	2.7	Citroën	12.0
Vauxhall	11.2	Vauxhall	9.0	Total BMC	46.8	Chrysler (UK)	
Rootes	10.9	Rover and Jaguar	3.2	Ford	28.4	Vauxhall	9.0
Rover	2.7	Others	0.8	Rootes	11.7	Datsun	4.0
Singer	2.1			Vauxhall	12.7	Renault	4.0
Jaguar	1.6			Others	0.4	Volkswagen	4.0
Others	2.8					Others	9.0

Sources: Drawn from data in D.G. Rhys, *The Motor Industry* (London, 1972), pp. 19, 20 and 67; P.J.S. Dunnet, *The Decline of the British Motor Industry* (London, 1980), p. 159.

concentration of production. There was a strong economic logic behind the search for merger opportunities in the British motor industry since no domestic manufacturer in the 1950s and 1960s approached the 500,000 units per annum output level at which significant scale economies occurred.

In contrast, the development of the American industry differed in both the nature and the timing of its changes. The giant corporations of GM, Ford and Chrysler had become all-powerful in the sector from the 1920s and from the 2,000 car firms of the early twentieth century only four so-called 'independents' remained in 1950: Nash, Hudson, Studebaker and Packard. In the following decade the independents prospered briefly with their distinctive smaller 'compact' cars, but they could never compete on cost with the three larger corporations, which enjoyed scale economies at every stage from research through to manufacture and distribution. By 1967 American Motors' (the product of a Nash–Hudson merger) market share had been severely reduced and Studebaker–Packard's wiped out as GM, Ford and Chrysler moved to dominate the market. Compared to Britain the vast American market offered benefits of marked scale economies[7] which in turn led the US manufacturers to adopt high levels of vertical integration for the production of components and materials.

The 1960s and 1970s form a third discrete period in the development of the British industry. This is evident from the sector's output record (see table 2). There was general growth until the 1960s with two major interruptions around each World War and four sharp breaks between 1945 and 1970. Each postwar drop in production has been related to government intervention to control demand in the economy. Expansion of output was continuous from 1946 to 1955 thanks to excess domestic demand. An engineer from Ford noted that 'business ... was obtained without being sought', in comparison to the interwar years when 'it had to be fought for every inch of the way'. The increase in car production by a factor of over four during the 1950s went in parallel with a quickened pace of technological change following the introduction of automatic transfer machinery for engine and sub-assembly.[8]

By contrast the later 1960s and 1970s proved more difficult. A sequence of critical events led to major market upheavals. Building on

---

[7] *The Ford Motor Company*, CIS Anti-Report no. 20, n.d., p. 2; C. Edwards, *The Dynamics of the United States Automobile Industry* (Columbia, 1965), p. 274.

[8] C. Malcolm, 'Note on the UK car industry', mimeo BLMC, n.d., p. 2; J. McNulty, 'A quarter-century of Ford production methods', *Institution of Production Engineers, Proceedings* (1954–5), p. 243.

Table 2 Output growth of the UK auto industry 1946–1970

Year	Output (thousands)	Year	Output (thousands)	Year	Output (thousands)
1946	219.2	1956	707.6	1966	1,603.7
1947	287.0	1957	860.8	1967	1,552.0
1948	334.8	1958	1,051.6	1968	1,815.9
1949	412.3	1959	1,189.9	1969	1,717.1
1950	522.5	1960	1,352.7	1970	1,641.0
1951	475.9	1961	1,004.0		
1952	448.0	1962	1,249.4		
1953	594.0	1963	1,607.9		
1954	769.2	1964	1,867.6		
1955	897.6	1965	1,722.0		

Source: Drawn from data in C. Malcolm, 'Note on the UK car industry', mimeo BLMC (n.d.), table 1.

the expansionary force of the previous period, world output in the 1960s accounted for almost 60 per cent of the total number of vehicles ever built. Productive capacity in the Western economies outpaced demand, however, as the abolition of tariffs in Europe led to a slowing down in market growth. Fierce competition ensued. The trauma of the 1970s for British manufacturers was occasioned by the twin oil price rises of 1973–4 and 1979. Besides a general decrease in production by all car-makers, the effect of the four-fold increase in oil prices led to a shift to smaller cars. Adjustment to reduced demand was difficult for all manufacturers since the orthodox assumption in 1973 had been that 1974 would be a record year for sales. The long-term, deep-seated problems of the British car industry – low productivity, high density of conflict and a poor managerial performance – arrested growth in the face of slack demand and a tendency towards saturation in the European markets.[9]

[9] W. Lewchuk, 'The British motor vehicle industry 1896–1982: the roots of decline', mimeo McMaster University, 1983; G. Bloomfield, *The World Automotive Industry* (London, 1978), p. 327.

International change: the 1970s

The 1970s stand apart in the history of the British automobile industry for one principal reason: the extent of the changes occurring on an international scale. By the end of the decade all domestic car industries were forced to compete on a worldwide basis and in 1980 a leading commentator on the industry observed how 'the global demand/supply picture for motor vehicles is crucially important in understanding the local changes which have taken place'.[10] The car industry of the 1970s had reached a state of maturity in the Western economies. Broadly speaking, maturity means a small number of producers using standard, intensive mechanization based on assembly line production for a market with decreasing opportunities for new sources of demand.

In growth terms, the 1970s represented a major discontinuity. By 1978–9 the rate of growth in world production was 2.8 per cent per annum, while it had been 6.8 per cent for the period 1960–73. The two acute recessions of 1973–5 and 1979–81 were touched off not by the oil shocks alone but by a combination of pressures. A general slackening of demand was expected by some experts owing to the gradual saturation of markets in the main industrial economies. International inflation rates and the rising real cost of motoring further decreased consumer expenditure on cars. Chanaron has therefore concluded that the recession in the automobile industry was 'essentially a normal cyclical down-swing characteristic of markets close to saturation. The oil crisis merely accentuated the decline'.[11] In 1978–9 production dropped by 6 per cent in France, 8 per cent in Germany and 12 per cent in Britain.

The most important markets by the 1970s in world terms were the USA, Western Europe and Japan. The world automobile industries had been overshadowed by America from 1945, when it produced three-quarters of the total vehicles; by the 1970s that proportion had been reduced to under one-third as the other major producer nations developed. Western Europe's market was created and sustained by the increase in world demand in the 1950s and 1960s. The integration of the market was enhanced by a series of political acts. Japan is the most recent market and industry to have reached maturity and by the 1970s

---

[10] M. Hincks-Edwards, Foreword to K.N. Bhaskar, *The Future of the World Motor Industry* (London, 1980).

[11] J. Chanaron in J.P. Bardou, J.J. Chanaron, P. Friedenson and J.M. Laux, *The Automobile Revolution: the Impact of an Industry* (Chapel Hill, 1982), p. 46.

Table 3 Shares of world car production and trade

	Production (millions)				Shares of world exports (%)		
	1960	1970	1978		1963	1973	1978
N. America	6.1	9.2	10.3	N. America	21	26	23
W. Europe	4.8	10.0	11.3	W. Europe	68	54	51
Japan	0.1	3.1	6.0	Japan	3	12	19
Latin America	0.1	0.6	1.2				
E. Europe	0.3	0.8	2.3				
Other	0.2	0.4	0.7				
Total	11.6	24.1	31.8				

Source: Drawn from data in D. Jones, *Maturity and Crisis in the European Car Industry: Structural Change and Public Policy* (Brighton, 1983), tables 3 and 8.

accounted for over 20 per cent of world sales.[12] The dominance of the three mature market areas and their producers is shown in table 3.

The development of the British car industry becomes particularly significant in the context of the mature market conditions of the 1970s. This is especially true with regard to the degree of integration of the leading vehicle producers. When British Leyland was formed in 1968 there was already a marked increase in the pace of rationalization in the European industry, following a spate of mergers and co-operative agreements. These changes were induced by the prospect of a unified market covering the EEC as tariffs were lowered. By 1970, therefore, three-quarters of the Western European car market was attributable to seven main firms. The crisis of the 1970s added further impetus to European integration among the car companies since, by way of merger and association, they could not only gain the benefits of scale economies but also acquire the capital required for the sharply rising costs of research and development and automation – in other words technological innovation. In 1974 Wells identified fourteen major

---

[12] Bloomfield, *World Automotive Industry*, p. 3; Bhaskar, *Future of World Motor Industry*, pp. 11 and 22.

linkages between the main car firms.[13] The result of these moves was a new profile for Western European car manufacturers, in which Britain found a place as a second rank producer (see table 4).

It was this period, when Britain fell from being a first rank international producer, that had severe consequences for innovation in the automobile industry. The industry entered the 1970s with a collection of inherited disadvantages which by now are well known. The comparative lateness of company integration and the continuing problem of internal reorganization within the context of generally slow growth resulted in an industry with 'too many plants producing too few cars'. Product ranges were misshapen and dated; low investment in the preceding 20 years coupled with the inflation of the early 1970s meant that the British companies lost even their domestic cost advantages. Productivity in absolute terms remained lower and its growth slower than for Britain's European counterparts. Production, management and profitability weaknesses offset the potential benefits available to foreign-owned companies located in Britain. British car-makers faced the difficulties of the 1970s just as they were attempting to restructure and retrench.[14] In the international organization of production, the UK industry fell drastically behind and could not therefore approach the scale economies of its foreign competitors.

Business strategy, change and the automobile manufacturers

In America, business strategists have used the automobile industry for both research purposes and models.[15] In Britain, a comparable relationship and body of knowledge have been slower to develop. The lateness of the industry's rationalization and the resulting problems have meant that the industry was unlikely to be the exemplar of strategic thinking in the way that GM became in the USA. On the contrary, it has taken considerable effort to uncover the business policies pursued within the UK industry. None the less, this task is doubly important since the role of strategy in understanding innovation has been generally neglected in the case of the British car-

[13] L.T. Wells, 'Automobiles', in R. Vernon (ed.), *Big Business and the State: Changing Relations in Western Europe* (London, 1974), p. 246; Automotive Industry Data, *Joint Ventures* (London, 1983).
[14] D. Jones, *Maturity and Crisis in the European Car Industry: Structural Change and Public Policy* (Brighton, 1983), pp. 12ff., 26.
[15] See, for example, A.D. Chandler, *Strategy and Structure: Chapters in the History of the American Enterprise* (Cambridge, Mass., 1962).

Table 4 Major car manufacturers and car-producing countries in 1981

Car manufacturers	Thousands	% of world output	Car-producing countries	Thousands	% of world output
General Motors (USA)	3,904	14.3	Japan	6,974	25.5
Toyota (Japan)	2,248	8.2	USA	6,253	22.9
Nissan (Japan)	1,864	6.8	West Germany	3,578	13.1
Ford (USA)	1,320	4.8	France	2,612	9.6
Renault (France)	1,294	4.7	USSR	1,350	4.9
Volkswagen (West Germany)	1,151	4.2	Italy	1,257	4.6
FIAT (Italy)	878	3.2	United Kingdom	955	3.5
Honda (Japan)	852	3.1	Spain	855	3.1
Mazda (Japan)	841	3.1	Canada	802	2.9
Lada (USSR)	830	3.0	Brazil	593	2.2
Opel (West Germany)	810	3.0	Mexico	355	1.3
Chrysler (USA)	749	2.7	Poland	295	1.1
Mitsubishi (Japan)	607	2.2	Yugoslavia	268	1.0
Peugeot (France)	569	2.1	Sweden	250	0.9
Citroën (France)	534	2.0	Australia	214	0.8
Ford (West Germany)	487	1.8	Czechoslovakia	178	0.7
General Motors (Canada)	478	1.7	East Germany	177	0.6
Daimler-Benz (West Germany)	449	1.6	Argentina	139	0.5
BL (United Kingdom)	413	1.5	Netherlands	78	0.3
Ford (United Kingdom)	342	1.3	Romania	70	0.3

Source: Drawn from data in Sinclair, *The World Car Industry*.

makers.[16] The main concern here is to outline the characteristics of the strategies used within the UK industry (as firms sought to match market, product and production) and thereby to assess their influence on the pattern of innovation.

It is widely recognized that Henry Ford's motor company was, from the 1900s onward, responsible for many of the basic production principles for automobiles as well as other manufacturing industries. His extension of the moving assembly line principle and the use of satellite plants brought about a breathtaking reduction from 15 to 1.5 in the manhours required to assemble a car. The main feature of the Ford approach was its obsessive concern with product and production standardization rather than customer requirements. Henry Ford was an engineer not a salesman[17] and, in spite of remarkable sales figures of 199,000 in 1913 and 240,000 *per month* in November 1922, Ford's aversion to marketing enabled the rival General Motors to take 28 per cent of the US market by 1926 and 60 per cent by the 1930s – a lead which GM has retained ever since. It was only after the Second World War that Henry Ford II relinquished this narrow stance and began to recast Ford in the mould of GM, largely by recruiting their management.[18]

Ford's routinization of production through the use of specialized machine tools and the act of moving the work to the worker via an assembly line have been called the 'secret of mass production' which 'suddenly made intelligible the major themes of a century of industrialization'.[19] However, the other giant car producer, General Motors, was responsible for introducing different but equally fundamental elements of strategy and organization in the automobile industry and elsewhere. Alfred Sloan, president of GM from 1923, arguably shaped the motor industry more than Ford and has been described as 'the inventor of most of the ideas of product planning, market analysis and corporate strategy which now play such a big part in the success or failure of a motor company'.[20]

In the 1920s GM was a corporation of seven companies making ten different types of car: each company had its own product, price and

---

[16] Lewchuk, 'Roots of decline', passim; CIS Anti-Report no. 5, *British Leyland: The Beginning of the End?*, n.d.
[17] R. Wild, 'The origins and development of flow-line production', *Industrial Archaeology*, 2 (1974), pp. 43–55.
[18] J. Ensor, *The Motor Industry* (London, 1971), p. 12; CIS, *The Ford Motor Company*, p. 54.
[19] C. Sabel, *Work and Politics* (Cambridge, 1982), pp. 29, 32–4, 195 and 201–2.
[20] Ensor, *Motor Industry*, pp. 12–13 and 107.

sales policy. Only two of the companies made a profit. Sloan's answer was to construct a range of cars (one car for each price/quality section of the market) which would be profitable. Each marketed segment was defined so that it was large enough in demand terms to support mass production.

What is often overlooked is that 'Sloanism' was based on the combination of Ford's mass production techniques and novel forms of company organization and direction. GM was reformed on a federal structure, in which corporate executives evaluated and controlled a set of competing divisions within strict profit measures. Each division shared the engineering, planning and financial expertise of the whole corporation. The critical change was to combine a decentralized, multi-divisional organization structure with central, strategic decision-makers who exercised supreme authority over long-term goals and planning. By using this combination GM was able to plan for potential demand and adjust their operations so successfully.[21]

A major yet unnoticed aspect of the success of Ford and GM in harnessing mass production to meet the demand of the world's first mass market was the low emphasis given to innovation. The 'dominant designs' in product, process and work organization were established by the 1940s and remained unchanged: only small revisions occurred until 1970. This is not to argue that the two corporations had reached a point of stasis, rather that change was contemplated judiciously, and so was less risky and more controllable.[22] Sloan saw caution as the watchword: 'The policy we said was valid if our cars were at least equal in design to the best of our competitors' grade, so that it was not necessary to lead in design or run the risk of untried experiment.[23]

Annual model 'facelifts' or minor changes became the norm. Product and process change was not led by research but taken up only insofar as it fitted 'the broad aims of the enterprise'. It became axiomatic that all aspects of design should 'depart only marginally from existing well-proven concepts'.

In Britain, by contrast, the combination of a fragmented market with slowly emerging mass demand and the absence of a highly developed managerial science did not lead to a set of clearly defined

---

[21] A. Chandler, *Giant Enterprise: Ford, General Motors and the Motor Industry* (New York, 1964); M. Jellnek, *Institutionalizing Innovation: a Study of Organizational Learning Systems* (New York, 1979), p. 43.

[22] W. Abernathy, *The Productivity Dilemma: Roadblock to Innovation in the Automobile Industry* (Baltimore, 1978), pp. 32–7.

[23] Ibid., p. 34.

business strategies among the car companies. Not unlike the rest of British industry, the car-makers have been generally slow to articulate or codify their methods of relating market and product. Mainly empirical, eclectic and short-term approaches predominated[24] and the more abstract and theoretically informed policies of the US corporations had no real counterparts on this side of the Atlantic until the late 1960s.

Here there is only space to highlight the main features of the strategies of the UK producers, in particular of those that eventually made up British Leyland. As has been noted, the British car market remained intensely competitive among a larger number of firms than was the case for America and for longer into the twentieth century. The 1930s witnessed the UK car companies making real annual model changes (not the yearly restyling of Ford and GM) each autumn; this accentuated the seasonal pattern of demand and led to severe fluctuations in output and employment. Long-term planning was therefore difficult and given little chance to develop. Only at Morris in the late 1930s and again after the Second World War was it decided to produce cars over a set number of years with fewer technological changes to achieve more continuous production.[25]

In spite of these generally under-developed strategies certain trends within the car companies' policies stand out. This is the most noticeable in the priority given to the product, production and work organization aspects, especially after 1945. For over two decades major weaknesses in production and marketing were left untouched, only to be cruelly exposed by the advent of the heightened competition of the 1970s.[26] Strengths in vehicle technology and engineering were often nullified commercially by poor standards of factory production and flimsy market analysis. The mergers of Austin and Morris led to no effective rationalization of the constituent companies; this meant that such imbalances were left untouched, which further retarded the growth of a common strategy. Among the BLMC companies, the Leyland (commercial vehicle) Company remained heavily sales-oriented while BMC was run by engineering experts. The result was that long-term product planning simply did not exist. The weakness in planning and marketing especially meant that in 1966 a misjudgement of demand by BMC produced a crisis that left the company hopelessly

[24] S. Pollard, *The Genesis of Modern Management* (London, 1965), p. 250; J.A. Merkle, *Management and Ideology* (London, 1980), p. 209.

[25] Rhys, *Motor Industry*, p. 17.

[26] Interview with senior product planner Austin-Rover, 1984.

over-committed on its production schedule.[27]

The British-based but American-owned Ford UK offered a very different type of strategic thinking. Like its American parent, before the 1960s Ford avoided product innovations on cost grounds and instead concentrated on using conventional engineering in conjunction with their highly developed volume production in order to produce cheap, essentially family cars. In the 1960s Ford management was regarded as the best 'all-round talent' in the industry, stemming from its use of graduate recruits against the traditional use of management recruits from the shopfloor.

Ford's marketing expertise was such that it consistently assessed demand trends more accurately than its competitors. The reorganization of Ford on a European scale during the same period, with a unified product plan for the whole continent, enabled the company to halve design time and achieve flexibility of production flow across national boundaries, and facilitated the successful transformation of the firm's product policy.[28] The emphasis changed from a total reliance on cheap models to a product range giving full market spread, the chance to optimize sales and profits in each sector of the market, and the opportunity to exploit the Sloan-inspired phenomenon of 'trading-up' by customers.

A major contextual influence on the car firms' business strategies and indeed on their ability to innovate has been the actions of the state. The record of government intervention in the car industry in Britain has been mixed. Some economists, such as Derek Aldcroft, have argued that the motor industry in the interwar period contributed significantly to the transformation of the economy. Others point to the failure of government policy to encourage the industry and optimise resource allocation.[29] Dunnett shows how in the postwar era *ad hoc* intervention regarding export levels and domestic demand had a deleterious effect on company planning and investment. Yet there is no statistical evidence for greater instability of demand in the UK than elsewhere. The frequent inability of the British automobile firms to exploit expansion in demand lessens the supposed impact attributed to government economic management.[30]

---

[27] Ensor, *Motor Industry*, p. 117.

[28] McNulty, 'Ford production', p. 247; Ensor, *Motor Industry*, pp. 84–94.

[29] D.H. Aldcroft, *The Inter-War Economy: Britain 1919–1939* (London, 1970), p. 187; Foreman-Peck, 'Market failure', pp. 286–7.

[30] Jones, *Maturity and Crisis*, p. 50; P.J.S. Dunnet, *The Decline of the British Motor Industry* (London, 1980), p. 121.

In comparison to the interventions of other European governments, notably Germany, the main features of British government intervention have been the lack of any coherent, long-term objectives for the industry, and the lack of consistency. During the 1960s and 1970s successive administrations had to learn two main lessons. First, there was a 'coming to terms' with the growing disparity between the performance of the UK car industry and the levels achieved by its competitors. Second, the inability of both the firms and the government to understand the depth of the problem for so long meant that the government's ability to assist the industry during the troubled 1970s was tightly constrained. At a time when worldwide changes required the British industry to innovate in almost every aspect of its planning and operations, the lack of a well-tried mechanism for developing appropriate government–industry strategic policy was glaringly apparent.[31] One acute example was the way the Labour administration energetically supported the formation of BL yet was unable to identify or advise on the innovations in business planning or organization which the new corporation required.

Car work

Descriptions of work in the car industry by social scientists have provided some of the most influential images of modern labour, and writers from a variety of backgrounds have also attempted to summarize changes in the organization of work across the industry. While some of these models of innovation in work organization have been widely accepted, they are of limited use in understanding the pattern of change in the British case. An examination of the inter-related shifts in the production process and methods of work shows why.

The pervading impression of the nature of work in the car industry was established by one of the earliest and most famous portrayals, Walker and Guest's study of American car plants in 1949 and 1952, which produced the following conclusion: 'the nature of the work itself, the fractionated repetitive conveyor pacing, profoundly affected all aspects of the quality of their working lives ... there was virtually no participation of the workers themselves in decisions affecting their working lives.'[32] More recently Guest has stated that, except for

---

[31] Wells, 'Automobiles', pp. 242 and passim; Jones, *Maturity and Crisis*, p. 50.

[32] For an overview of this research see R.H. Guest, 'Organizational democracy and the quality of work life: the man on the assembly line', in C. Crouch and F. Heller, *Organization Democracy and Political Processes* (London, 1983), pp. 139 and passim.

Sweden, 'there has been almost no fundamental changes [sic] in the basic structure of jobs on motor assembly lines anywhere in the world. They remain highly fractionated and conveyor technology still dominated [sic] the nature of work.'[33] A substantial body of literature supports these conclusions.[34] Yet on close inspection the orthodoxy fails to account for the changing nature of work in the UK car industry, because it ignores the key differences in strategic thinking and production history which have been outlined in previous sections.

Detailed accounts of work in British car factories reveal the differentiated experience of work, the variable impact of technological change and the ability of workers to control important aspects of their tasks and work routines. Warnings have been rightly issued against exaggerating the strength of union power in the industry given the late development of widespread organization in the 1950s. Yet a common theme emerges, from the studies by Lewchuk to the case-studies of Friedman, Clack or Melman and the personal accounts of Exell.[35] They all agree that the slowness of British car companies to adopt capital-intensive production techniques and the reliance more on payment-by-result systems to maintain effort and output made for weak managerial control over production. As Zeitlin puts it, this 'created a space' within which worker job-control could exist and expand in the circumstances of both the late 1940s and more especially of the 1960s. The comparative technological backwardness of the industry did not make for a uniformly unskilled labour force: gradations of skill abounded. The fragmented production sequence gave many groups of workers the ability to halt the process. Management recognized their importance and these groups developed their own traditions of independent organization.[36]

Bardou and others[37] have constructed an historical profile of the organization of work in Western Europe, but their model is of limited

[33] Guest, 'Man on the assembly line', p. 152.

[34] For the USA see E. Chinoy, *Automobile Workers and the American Dream* (New York, 1955); for the UK see H. Beynon, *Working for Ford* (Harmondsworth, 1973).

[35] S. Tolliday, 'Trade unions and collective bargaining in British motor industry 1896–1970', mimeo Kings College Research Centre, Cambridge University, 1984; Lewchuk, 'Roots of decline'; A. Friedman, *Industry and Labour: Class Struggle at Work and Monopoly Capitalism* (London, 1977); G. Clack, *Industrial Relations in a British Car Factory* (London, 1967); S. Melman, *Decision-making and Productivity* (London, 1958); A. Exell, 'Morris Motors in the 1930s', *History Workshop Journal*, 6 (1978), pp. 52–78.

[36] J. Zeitlin, 'The emergence of shop steward organisation and job control in the British car industry: a review essay', *History Workshop Journal*, 6 (1980), pp. 119–37.

[37] Bardou et al., *Auto Revolution*, pp. 247ff.

application to Britain. Their central argument is that the advent of mass production through the imitation of US management methods from 1945 onwards led to new attitudes towards work. In the absence of identification with the new regime, workers' absenteeism and labour turnover increased. Antipathy to the 'new industrial organization' resulted in unprecedented labour action in the late 1960s and 1970s. The so-called 'new workers' (especially migrants) found no adequate expression of their views in the established unions and the result was a host of 'wildcat', short-term workplace unrest, seen in its most pronounced form in the 1970–3 period.

While many of the features of this scheme of industrial work organization in the European car industry of the postwar era do find echoes in Britain, they are only faint. The UK industry deviated in a number of ways. The phenomenon of migrant or guest workers was not relevant.[38] It has already been shown how resistant British management was to US techniques, so that their adoption was slow and fitful. In Britain, it could be argued that spontaneous shopfloor action arose not so much as a response to far-reaching, management-led reorganization but in reaction to the frustrations generated by the industry's deep-seated problems outlined above – problems which were raised to a greater intensity by the mergers and higher competition of the 1965–75 period. In further contrast to Europe, the multi-union UK car industry and the absence of a national bargaining 'system' meant that the most frequently recorded disputes arose over wage structures and inter-union relations, not over new technology or methods as Bardou et al. maintain.[39]

Beyond industrial relations, the changes in production techniques also accounted for some of the key developments. By far the most penetrating analysis of the general process of change in car manufacture is provided by Abernathy, who observed how in the early part of the century fierce competition in the industry led to rapid technological progress. During the industry's growth from the 1920s, radical product and process innovation was replaced by the gradual dominance of complex mass production machinery with its attendant standardization resulting in incremental change. There appeared to be a paradox: the conditions necessary for rapid innovation were the opposite of those that supported high production efficiency. The

---

[38] D. Lyddon, 'Workplace organisation in the British car industry', *History Workshop Journal*, 15 (1983), pp. 131–40.

[39] H.A. Turner, G. Clack and G. Roberts, *Labour Relations in the Motor Industry: a Study of Industrial Unrest and an International Comparison* (1967), p. 290.

characteristics of product and process converged as the car-makers sought stability in incrementalism and caution.[40]

Abernath's survey of the reconstruction of the UK car industry emphasizes the different innovation trajectories taken by Britain and the USA. His account is heavily based on the American companies; consequently he has been able to relate such changes to a set of clearly defined corporate strategies regarding manufacturing policy. By contrast, the UK producers pursued much less straightforward approaches. The rationalization of the British car-makers was frustrated by the slow emergence of a mass market as well as by the very different selling seasons in the UK. Unpredictable demand was simply not conducive to the increasingly specialized plant or methods needed for true mass production; this is confirmed by detailed Anglo-American comparisons carried out by UK engineers in the 1940s and a major review of Austin techniques in 1947, which showed that US methods were inappropriate for their needs.[41]

In the face of the so-called 'dominant design' in plant across the world after the 1920s, British evidence clearly indicates that the fateful combination of low investment and the persistence of inherited layouts meant that domestic firms did not possess highly specialized and integrated plants even in the 1960s. The over-reaction of mechanical engineers to what was essentially a very limited introduction of automation using transfer machines in the 1950s is indicative of the retarded development of production technology in the British car industry.[42]

While the US auto manufacturers in the 1960s and 1970s found themselves in a crisis of productivity because of product and production features which compared unfavourably with the emergent producers of the Far East, the UK trauma stemmed from an inability to rationalize and meet the efficiency levels set by those US companies in the inter- and immediate postwar periods.[43] In contrast to the Swedish and Italian automobile industries, which adopted new methods of work organization based on group working and early robot technology in the 1970s, UK companies fell victim to their accumulated problems. Rather than matching the innovations of Volvo or Fiat, the

---

[40] Abernathy, *Productivity Dilemma*, pp. 13–36.

[41] F.G. Woollard, 'Engine assembly. The system and layout employed by the Austin Motor Co. Ltd.', *Automobile Engineer*, 37 (1947), pp. 90–7.

[42] H.J. Graves, 'An outline of B.M.C. development in the field of automation', *Journal of the Institution of Production Engineers*, 36 (1957), pp. 18–20.

[43] Bhaskar, *Future of World Motor Industry*, p. 260.

British industry was still struggling against the historical inertia of low levels of company integration, outdated plant, fragmented bargaining structures and restricted market outlets.[44]

Conclusion

This brief overview of innovation in the British automobile industry has been able to do no more than sketch out the main outlines. However, a number of points are worth highlighting and deserve closer inspection not only by historians but also by those who are grappling with the so-called 'technological ferment' and 'new' competition of the 1980s.

Perhaps the most important features are those which distinguish the British from the US experience. Without doubt Britain, like her continental neighbours, mirrored some of the broad aspects of the American pattern of innovation, yet it is the decisive differences in the areas of timing, market and strategy which stand out.

The divergence in the long-term profile of the UK from the American car industry is the most easily identified. While the US producers had reached a state of business concentration and technological maturity by the 1940s, Britain was still in adolescence in the late 1960s. The crisis of the 1970s had a very contrasting meaning for them. Apart from the opposite effect of the two World Wars on the British and American industries, and the significant distortion of the UK sector's long-term growth, the dissimilarities in market structure grew more pronounced. In Britain, the absence of an early mass demand, coupled with the persistence of a more stratified set of consumers (the basis after all of Rootes sales until the 1960s) meant that the process of company concentration was slow and uneven in comparison to that of the USA. The word uneven is used with care: while Ford and GM witnessed the contraction of their competitors at home, the expansion of the twin giants abroad increased the number of producers in the UK.

Given these structural contrasts, the distinctive business strategies of the UK companies take the Anglo-American divergence even further. While commentators in the 1970s bemoaned the lack of innovative activity in almost every aspect of the US firms, in Britain the level of innovation was rising markedly as deeply embedded, anachronistic

---

[44] Work Research Unit, *Developments in the Quality of Working Life in the Motor Industry* (London, 1978).

production and work organization policies were overhauled – the introduction of measured day-work is a good example. If the American profile of change can be portrayed as a graph line showing intense activity around the first three decades of the century followed by a slowly declining curve, then the UK car industry profile could not be more different. The intersection of a unique company and market structure with the constraints imposed by the character of state intervention, or non-intervention, encouraged companies to adopt policies which did not lead to incremental change and overwhelming caution. The high market value of British product engineers and car designers worldwide was based on their innovative capacity.

These features of the life course of the British car industry are vital to an understanding of its distinctive position compared to not only the USA but also the rest of Western Europe. There are, however, two further aspects to the profile of change which are of importance, timescale and the inter-relation of the key dimensions of innovation. The question of time is the more straightforward.[45] The preceding account makes it clear that is is impossible to understand the profundity of the crisis which the UK car industry faced in the 1970s and 1980s without an historical approach. In terms of product technology the industry was relatively well equipped to meet the requirements of greater sophistication and increased levels of vehicle innovation. It was in the areas of manufacture and management that the car companies were less well endowed. The crucial aspect of these weaknesses was their longevity. As a result, companies which had based their postwar development on domestic replacement demand and lifted Commonwealth or related exports had not found it necessary to promote the marketing or automated production techniques of their continental rivals. The length of the postwar replacement boom and the extended growth of demand through the 1950s and 1960s meant that these deficiencies became embedded across time: the impact of fierce competition following the oil price rises was all the more difficult to face.

The second feature of this account of the British car industry is also of wider significance. Although this is only an outline of a single sector of the economy, the manner of reconstructing the profile of change may help in understanding industrial innovation in other sectors. Change can be explained if we acknowledge its inter-related character. Innovation cannot refer solely to technological change, it must encom-

---

[45] R. Whipp, 'A time to every purpose: an essay on time and work', in P. Joyce, *The Historical Meanings of Work* (Cambridge, 1987), pp. 210–36.

pass the linked shifts in product, production and work organization. It is now more common to relate product and process changes when prompted by the present-day example of computer-aided engineering and computer-integrated manufacture; it is even more important, though, to see these linked conceptually in historical studies of innovation. The managements of the car-producers in the 1980s have been forced to face up to the triangular dimensions of industrial change, and so must historians of innovation if they are to contribute to the current debates over the future of the UK and world automobile industries.

8
Technology and the Export of Industrial Culture: Problems of the German–American Relationship 1900–1960

Volker R. Berghahn

While much of today's business history continues to be concerned with the study of individual firms, it is possible to discern among the practitioners of this genre of historical writing a renewed interest in larger national and international developments, as well as in their impact on company culture in the narrow sense. This is not to say that research into entrepreneurial attitudes and historical traditions within the larger context of national experience is a recent phenomenon. On the contrary, the cultural perspective experienced a major boost in the 1950s.[1] But it was different from that of the 1980s. Inspired by modernization theory, its main concerns were with identifying the peculiarities of a particular national business milieu and explaining why certain countries developed in different directions while allegedly displaying a greater or lesser dynamism as they evolved towards an advanced industrial economy and society. The debate about the supposedly peculiar path of French industrialization in the nineteenth and twentieth centuries is probably the best-known case in point. In the meantime many of the assumptions and data of this type of

---

[1] This approach was particularly influential in the United States and spearheaded by scholars like R. Bendix, R. Cameron, F. Redlich and G. Almond.

research were challenged, as perspectives began to change.[2]

One aspect of which we have recently become more acutely aware is that, just as individual companies operate within a larger national industrial culture, a national industrial system cannot be seen in isolation from those of other countries; it is exposed to foreign business practices and behavioural traditions and responds to them, especially to those which radiate from a country that enjoys a hegemonic position in the world economy. Indeed the problem of hegemony has become an important field of research, as scholars began systematically to assess the superior weight exerted by the United States in the postwar world, as well as the gradual erosion of that superiority since the late 1960s. This field opened up the possibility for explicit and direct international comparisons and hence for raising different sets of questions from those of the 1950s, i.e. how American industrial culture influenced the milieu and traditions of other capitalist economies and how those nations reacted not merely to the introduction of American machinery, but also to exports of an organizational and cultural kind. Once the problem had been posed in this way, it became virtually inevitable for the interested business historian to ask a further question, i.e. how far back we can trace the roots of the American impact and what may be learned from the interwar or even the pre-1914 experience. And once we add this historical dimension, it is in turn surprising to find how much the contemporary debate in interwar Europe was preoccupied with the 'American question'.[3] As the Italian marxist Antonio Gramsci put it programmatically in the late 1920s: 'The European reaction to Americanism ... must be examined attentively. Analysis of it will provide more than one element necessary for understanding the present situation of a series of states of the old continent and the political events of the postwar period.'[4]

It would be interesting to pursue why the importance of this statement and of similar insights was lost from sight until the 1970s. No doubt the developments of the 1930s and early 1940s put a heavy damper on the further exploration of Gramsci's proposal. It was a time of nationalist introspection. There existed a widespread impression that the USA had never really entered the interwar world and had

[2] See, e.g., P. O'Brien and C. Keyder, *Economic Growth in Britain and France, 1780–1914. Two Paths to the Twentieth Century* (London, 1978).

[3] See, above all, F. Costigliola, *Awkward Dominion. American Political, Economic, and Cultural Relations with Europe, 1919–1933* (Ithaca, 1984).

[4] Q. Boare and G.N. Smith (eds), *Selections from the Prison Notebooks of A. Gramsci* (London, 1971), p. 281.

persisted in its isolationism. So, in the light of these assumptions, what hegemonic impact of American industry on that of Europe was there to be studied? It was only through the pioneering research in the 1960s of scholars like W. Link that a fresh start was made.[5] He showed that, political isolationism and isolationist rhetoric notwithstanding, the USA became deeply involved in Europe economically and financially. Once the Dawes Plan had been ratified in 1924, American loans flowed across the Atlantic and American companies invested directly in Europe, establishing production bases in France, Britain, Germany and elsewhere. This kind of work by political and economic historians laid the groundwork for the business historian to return to the Gramsci quotation, to look at what had been written on the subject during the 1920s and to link up with the research of sociologists like H. Hartmann. His early work had been very much within the framework of early post-1945 studies on divergent business cultures. Thus, in 1959, Hartmann published his *Authority and Organization in German Management*,[6] in which he investigated the peculiarly authoritarian and elitist attitudes of the German business world. He found these attitudes to be alive in the 1950s and emphasized their deep-rootedness in the German industrial experience in the first half of this century. Insofar as Hartmann made international comparisons, they were more indirect and assumed American business culture to be less authoritarian and more democratic.

Subsequently, he extended the scope of his inquiry and in 1963 produced a monograph on *Amerikanische Firmen in Deutschland* (American Firms in Germany).[7] Here his main concern was with the export of American industrial culture and with its impact, primarily via American subsidiaries, on West Germany's industry. Hartmann proceeds by querying the then received view that technological exports pose fewer analytical problems than exports of organizations, attitudes and industrial ideologies. Indeed, to him it is misleading to differentiate between so-called instrumental exports, on the one hand, and cultural ones, on the other. Technology, he believes, is never transferred as such as another country. There are always values behind it which cannot be separated out. The adoption of foreign technologies also involves the taking-over of doctrines and attitudes of which the

---

[5] W. Link, *Die amerikanische Stabilisierungspolitik in Deutschland, 1921–1932* (Düsseldorf, 1970). See also the important recent study by W.C. McNeil, *American Money and the Weimar Republic* (New York, 1986).

[6] Princeton, 1959.

[7] H. Hartmann, *Amerikanische Firmen in Deutschland* (Cologne, 1963).

technologies are the tangible expression. This implies that the introduction of instrumental imports is anything but a simple process. What, according to Hartmann, must also be studied are the assumptions and mentalities of the receiving industrial culture, because these act as filters. These filters will test how far the proposed imports are compatible with indigenous traditions. In other words, there are potentially powerful impediments to the importation of foreign technologies and techniques, the analysis of which tells us something more generally about the character of the two industrial systems and the relationship between the hegemonic and the dependent power. If the exports originate from a non-hegemonic power, the impediments may be strong enough to block acceptance altogether. Exports from a hegemonic power are more difficult to resist and hence likely to lead to a more confused picture; they are likely to divide the indigenous business community into those entrepreneurs who believe the underlying value systems to be incompatible with one another and those who are more open to imports, either because they see adaptation to the hegemony as inevitable and desirable or because the imports contrast less sharply with their own attitudes and traditions. The result of such divisions is likely to be a pattern which reflects changing balances between divergent branches. The existence of varying 'filter strengths' even makes it conceivable that technologies alone may be adopted, e.g. in the process of a rationalization programme, with the unspoken assumptions successfully deleted, thus requiring a modification of Hartmann's hypothesis.

The questions concerning technology and the export of industrial culture which have been raised so far will now be further investigated and tested by reference to the case of Germany. The main testing ground will be the reception of American ideas about factory production and organization, as first propagated by the Taylorists and later by Fordism.

The general principles of F.W. Taylor's system and his engineering approach to industrial work organization have been examined many times.[8] It is widely accepted that he and his followers, in promoting the idea of greater productivity and efficiency in the machine age, were not merely concerned with time-and-motion studies and more rational shopfloor organization; they also made far-reaching assumptions about the nature of man and demanded a mental revolution by

---

[8] See, e.g., H. Braverman, *Labor and Monopoly Capital* (New York, 1974); S. Haber, *Efficiency and Uplift* (Chicago, 1964); L. Urwick and E.F.L. Brech, *The Making of Scientific Management*, 3 vols (London, 1945).

employers and employees alike. Taylorism was not a uniquely American phenomenon, and it is not the case that similar assumptions could not be found in Europe. However, there can be little doubt that the American situation of the turn of the century provided a particularly fertile ground for the development of Taylorism and Scientific Management: a highly mobile immigrant society with a large pool of unskilled non-unionized labour captivated by ideas about material self-improvement, a democratic populism and notions of Social Darwinist competition. There was also 'the commitment to technological efficiency and productivity' which 'pervaded almost the entire culture', whereas in Europe 'it appeared more selectively'.[9] Indeed, the situation of the societies and industries of Europe, with their old hierarchies and their traditionalist paternalism, was tangibly different from that of the USA. Still, this difference did not prevent many Europeans from developing a keen interest in that young nation across the Atlantic with its seemingly unlimited economic potential. Clearly, this was one of the major powers of the future. One began to take note.

In line with such perceptions, German diplomats and military officers began to study and to include the USA in their power-political and military-strategic considerations. Similarly industrialists and technical experts turned towards American industry, its products and technologies. At the 1900 World Exhibition in Paris, new steel-making technologies received much attention. Some observers even began to perceive an 'American danger' for the future of German industry, thus providing a further incentive for monitoring what was happening on the other side of the Atlantic.[10] It looked as if something like a 'technological gap' was developing between Germany and the USA in areas such as precision tool manufacture and standardized production techniques. The response by German business reflected the inchoate hegemonic relationship: soon experts could be seen travelling to the USA to gather first-hand information. Among them was Paul Möller, whose impressions of a seven-month visit first appeared in 1903 in the journal of the *Verband Deutscher Ingenieure* (VDI, Association of German Engineers). Inevitably these visitors also came across the publications of Taylor and his followers. In terms of our argument it would seem significant that Taylor's 1906 essay, 'On the Art of

---

[9] C.S. Maier, 'Between Taylorism and technocracy', *Journal of Contemporary History*, 2 (1970), p. 28.

[10] H. Homburg, 'Anfänge des Taylorsystems in Deutschland vor dem Ersten Weltkrieg', *Geschichte und Gesellschaft*, 2 (1978), pp. 173ff., also for references in the following text.

Cutting Metals', was picked up without longer delays, whereas his 'Shop Management' (1903) was translated into German only six years after its first publication and was received much more sceptically. Consequently it also took people longer to realize that Taylor offered more than more efficient shopfloor organization and was in fact dreaming of an entirely new industrial system and of a fundamental change in attitudes on the part of workers and entrepreneurs.

It was at this point that the question was bound to arise as to whether that whole edifice was suitable for use under German conditions. The above-mentioned filters became activated; critics and protagonists emerged. The debate saw a further intensification after 1912 when Taylor's *Principles of Scientific Management* appeared in German translation and reached its third impression by the spring of 1913. The annual general meeting of the VDI that same year spent much of its time discussing modern methods of industrial management and factory organization. By now Taylorism had ceased to be a topic reserved for argument among engineers and technical experts. Its wider social significance had been recognized, even if opinions remained divided over what to make of it. Thus when A. Wallichs, an engineering professor at Aachen Technical University, defended Taylorism against its critics in the liberal *Frankfurter Zeitung*, the paper added an editorial postscript stressing that the 'cultural consequences' of the new system were still completely obscure. Similar considerations apparently also motivated F. Neuhaus, the director general of the Borsign Works, and P. Perls, a Siemens-Schuckert director, who, though favourably disposed towards Taylorism, nevertheless were uneasy about a slavish imitation of the new system.

Consequently, as L. Burchardt has argued, Germany lagged far behind the United States in the discussion of Taylorism and its wider implications.[11] There was no organized movement similar to the American one and only a few individuals, such as Wichard von Moellendorff, built upon the larger assumptions and extolled the alleged positive effects for German society as a whole. Above all, according to Burchardt, German entrepreneurs, with a few marginal exceptions, never actually began to experiment with Taylorist methods. On this latter point H. Homburg has taken a different view.[12] She believes that the important electrical engineering firm of

---

[11] L. Burchardt, 'Technischer Fortschritt und sozialer Wandel. Das Beispiel der Taylorismus-Rezeption', in W. Treue (ed.), *Deutsche Technikgeschichte* (Göttingen, 1977), pp. 52–98.
[12] H. Homburg, 'Anfänge des Taylorsystems', pp. 180ff.

Robert Bosch at Stuttgart practised 'the beginnings of the Taylor System' in the decade before 1914. American ideas on modern factory organization were apparently first picked up by a number of Bosch managers from 1904 onwards, after they had been across the Atlantic. In 1913, then, H. Borst, a member of the Bosch board of directors, paid Taylor a longer personal visit and thenceforth propagated his ideas. Later Borst claimed that Bosch had arrived at new forms of management and organization independently of Taylorism and it may well be that the diffusion of ideas from across the Atlantic was more subtle and piecemeal. Nevertheless, by 1910 Bosch had certainly undergone major reorganizations in such areas as production and financial control.

Other recent research on the pre-1914 German car industry reinforces Homburg's work on Bosch.[13] According to P. Conrad, writing in the journal *Der Motorwagen* in 1905, 'Americanization advances in Germany at an accelerated pace'. This was true not merely of the latest machinery, which appeared in companies like Daimler, but also of changes that were made in the firm's organizational structure. However, ultimately it was not Daimler but Opel which became the pacemaker in the standardization movement and the introduction of American methods in metal engineering, first in the fields of sewing-machine and bicycle production and later in automobile manufacture. The principle of Opel's revised workshop organization soon also affected other parts of the company. Indeed, the ideas that Opel introduced in 1906 'coincided in many instances with proposals' which VDI member Ernst Valentin had published in *Der Motorwagen* in 1904 following his visit to the United States. Similarly Wilhelm Wenske, the technical director of Opel's *Arbeitsbüro*, became a student of Taylorism. In 1906 he joined the Executive Committee of the *Automobiltechnische Gesellschaft,* which organized regular lecture series on American ideas about factory management. In short, the rationalization movement appears to have advanced on a broader front and in particular began to affect the industries of the second industrial revolution, such as electrical engineering and automobiles.

One of the repercussions of these developments was changes in the structure of the workforce of these industries. As the Stuttgart firm Bosch introduced its rationalization programme, adopting stricter divisions of labour and repetitive machine-aided manufacture, the deskilling of blue-collar workers, which the critics of Taylorism had

---

[13] A. Kugler, 'Von der Werkstatt zum Fliessband', *Geschichte und Gesellschaft*, 3 (1987), pp. 304–39, also for references in the following text.

always predicted, set in. Meanwhile the pace of the manufacturing process also accelerated, and by 1909 the 'Bosch tempo' had become something like a derogatory term. At the same time the Bosch management began to erode the collective bargaining position which it had granted earlier to the *Deutscher Metallarbeiter-Verband* (DMV), the union of skilled metal workers. As a result, tensions continued to mount over several years until they finally erupted in a major strike in the summer of 1913.[14]

There are many similarities between what has been said so far about the German case and the development in France, thus pointing to a wider impact of American exports upon Europe. In France, too, Taylorism made its first impression on technical experts around the turn of the century. Interest in American ideas on scientific management grew rapidly in the years just before the First World War, although industrialists proved reluctant to apply them in their factories. Only French car manufacturers, and Renault in particular, began to experiment with Taylorist methods, but they 'were thwarted ultimately by French workers'.[15] As in Germany, 1912–13 saw a wave of 'strikes against Taylorism' when, as G.C. Humphreys added, the concept of

> management control over the work process and the subdivision of the production process [came into] direct conflict with traditional handicraft values of French machinists, who had maintained a significant degree of autonomy in French automobile factories. Thus it was the threat that Taylorism presented to the skilled workers' basic assumptions about work and their social status which forced the CGT (their union) to take its position against scientific management.

National and regional variations always granted, it seems that similar concerns were also at the heart of the resistance of the DMV and its members to Bosch's Taylorist innovations. The export of scientific management, once it had turned out to be more than a purely technical and organizational device, ran up against considerable obstacles under the conditions of ingrained European labour traditions and labour market structures. Cultural filters became operative. Of course, it is less surprising that workers and organized labour should oppose attempts to apply Taylorist methods in German factories. But the employers too put up resistance, even in industries like car manufacture

[14] H. Homburg, 'Anfänge des Taylorsystems', pp. 180ff.
[15] G.C. Humphreys, *Taylorism in France, 1904–1920* (New York, 1986), p. 247. See also P. Fridenson's writings on Renault.

which, if the American example provided any guide, were well suited for assembly-line production. Daimler presents a case in point: unlike Opel, this company shied away from volume car production because the owners deemed it to be 'incompatible with Daimler's self-image' and with the notion the company 'had of the value of its products'.[16] As a matter of fact, the management proudly juxtaposed its own high-quality cars for a luxury clientele with American cheap production for an anonymous market: 'Here (we do things) meticulously and thoroughly, over there *husch-husch-fertig* (skimping and rushing).' In line with this attitude, the company was not interested in cost and price reductions and declared with visible relief that 'over here we are still a long way away from the American situation where every clerk owns a car. With us the automobile is for the most part a vehicle for the better-off classes.'

It seems therefore that reactions to American technology and to Taylorism can also tell us something about cultural and social-structural differences between Germany and the USA and about attitudes to American imports among employers. Nor was it unexpected that resistance should continue and even sharpen during the interwar period. Although some union leaders and socialist intellectuals at times took a more positive or at least ambiguous stand towards American-inspired factory rationalization and efficiency drives, suspicions of the movement remained strong and, as we shall see later, were reinforced by the peculiar response of the employers.[17] Thus some of the more radical left-wingers saw increased rationalization and concentration as a step towards the gradual transformation of industrial capitalism into socialism, whereas others were prepared to test the promise of scientific management that better productivity would raise average living standards, resulting from larger wage packets and cheaper mass-produced goods in the shops.

If these latter hopes were sorely disappointed, this was not merely because of the difficulties which bedevilled the German national economy even during the more stable mid-1920s; it was also because of the behaviour of some of the employers and the ways in which they dealt with the new ideas that kept on coming into Germany from across the Atlantic. As we have already seen, a division began to develop in the entrepreneurial camp well before 1914 between a few

---

[16] A. Kugler, 'Von der Werkstatt zum Fliessband', pp. 314ff., also for references in the following text.

[17] See, e.g., H.A. Winkler, *Der Schein der Normalität. Arbeiter und Arbeiterbewegung in der Weimarer Republik* (Bonn, 1985), pp. 62ff., 466ff.

who were open to Taylorist methods and their broader ideological assumptions and a majority who discerned something like a culture clash and argued that German and American industrial traditions were basically incompatible. During the Wilhelmine period, hostility to American cultural exports was articulated by the *Centralverband der Deutschen Industrie* (CVDA), industry's peak association. As this organization was dominated by heavy industry, it seems that the main opposition came from the patriarchs in coal and steel, with a good deal of support from small and medium-sized companies in which the traditionalism and authoritarianism of the nineteenth century continued to hold sway. Here industrial relations remained 'punishment-centred' (A.W. Gouldner) and largely untouched by ideas about holding out material incentives to the workforce, not to mention arguments about any wider societal benefits.[18] The conservatives expected employees to be content with what was unilaterally and 'graciously' handed out to them. Bosch's early experiments with union recognition and collective bargaining were viewed with deep suspicion. When that company, as we have seen, ran into increasing trouble with the DMV and ultimately found itself embroiled in a major strike, the critics in the entrepreneurial camp responded with suitably subdued glee and *schadenfreude*.[19]

The experience of the First World War and of total industrial mobilization produced two significant changes affecting industrial relations in Germany: it increased the power of organized labour, enabling it to advance its claim to being the rightful bargaining partner of the employers; it also led to a growth of the indirect powers of the state over production and the market as well as of direct state intervention.[20] These developments widened existing divisions among the employers. By 1917–18, some wanted to roll back both the unions and the state and to return to the patriarchal conditions of the Wilhelmine age. Others proposed to reduce the powers of the state, but to retain the co-operation of organized labour. A third group finally argued that the economic and social problems at the end of four years of total war followed by defeat could only be resolved if the incipient wartime triangular relationship between employers, workers and the government was built upon and if a concerted effort was made

---

[18] V.R. Berghahn and D. Karsten, *Industrial Relations in West Germany* (Leamington Spa/New York, 1987), pp. 142ff.
[19] H. Homburg, 'Anfänge des Taylorsystems', p. 190.
[20] See G.D. Feldman, *Army, Industry and Labor in Germany, 1914–1918* (Princeton, 1966).

by all sides to use the interventionist measures developed during the war to put the national economy back on an even keel.[21]

Given the chaos of 1918–19, which had barely been brought under control when the country faced another total collapse in 1923, it is difficult to decide which of the three employers' positions gained the upper hand. It is easier to arrive at a cautious verdict as far as the years 1924–8 are concerned. This period witnessed the ephemeral ascendency of influential branches of industry which were more prepared to co-operate with the unions and the government on major issues of social and economic policy.[22] Significantly enough they tended to be the same branches which looked towards the United States, the hegemonic power across the Atlantic, for support and partnership. But, as before 1914, there were also the conservatives, especially in coal and steel, to be reckoned with. Indeed, as B. Weisbrod has argued, they may, though no longer dominant, even have retained sufficient 'veto power' to undermine and ultimately to scupper the plans of the more liberal sections of German industry.[23] But for the moment the actual outcome of this internal struggle over industry's position *vis-à-vis* the unions and the state is less important for our purposes than the effect which these particular developments had on technological and organizational imports from America. No doubt the course and outcome of the First World War had greatly strengthened the weight of the USA in the world economy. It is also clear that both Washington and the New England business establishment became deeply committed to a revival of the European economy as a precondition of political and social stabilization in the Western world. Recognizing Germany's superior, though under-developed, industrial potential, American politicians and businessmen came to regard her as the main country on which to concentrate their reconstruction efforts. As early as 1921, Norman Davis put this point to Secretary of State Hughes:

> Through the highly industrial development of Europe prior to the war, Germany has become the axis, and the rehabilitation of Europe and its continued prosperity is most dependent upon that of Germany. Unless Germany is at work and prosperous, France cannot be so and the prosperity of the entire world depends upon the capacity of industrial

---

[21] V.R. Berghahn and D. Karsten, *Industrial Relations in West Germany*, pp. 154ff.

[22] H.A. Winkler, *Der Schein der Normalität*, pp. 510ff.; D. Abraham, *The Collapse of the Weimar Republic*, 2nd edn (New York, 1986), pp. 106ff.

[23] B. Weisbrod, *Schwerindustrie in der Weimarer Republik* (Wuppertal, 1978).

Europe to produce and purchase. Into this enters the element of credit, and credit will not be forthcoming as long as there is no stability and confidence, and until the German reparation is settled constructively on a basis which will inspire confidence the credits necessary for the reestablishment of normal conditions will not be forthcoming.[24]

It took another four years and further upheaval before a reparations settlement, inspired and promoted by the USA, was reached. But once the Dawes Plan had been ratified in 1924, American loans became available. At the same time German trusts like IG Farben Chemicals and Zeiss Opticals sought the co-operation of American companies. Krupp and General Electric went into joint carbide production. Americans also invested directly in German firms or built up their own subsidiaries and production facilities in Germany. General Motors took control of Opel Cars in 1929, one of the main automobile manufacturers in Germany which, significantly enough, had begun to concentrate on the cheap volume-production end of the market. In 1927 Chrysler Motors opened an assembly plant in Berlin; Coca-Cola established bottling facilities at Essen in 1929. Thus the German economy became a 'penetrated system'.

Such developments were bound to give a boost to the old question of American technological exports, which formed the starting-point of our analysis. Of course, during the decade since 1914, Taylorist ideas had undergone a tangible metamorphosis. Above all, the ideas of Henry Ford had also arrived on the scene. Somewhat crudely put, the difference between Taylorism and Fordism was that whereas Taylor's remained an engineering approach to industrial production enlarged by quite utopian social claims, Ford's was that of a sober and hard-boiled *social* engineer who stressed the link between efficient and cheap mass production and mass consumption as a stabilizing device for a capitalist economy under conditions of political democracy.[25] Considering the increased weight of American industry in the world and the curiosity which its methods had aroused in Germany well before 1914, it is not surprising that leading German industrialists like W. Zangen or C. Duisberg and younger managers like W. Rohland, F. Berg and O.A. Friedrich were soon on their way to the industrial centres of North America. They were followed by delegations of trade unionists. Ford's factories in Michigan inevitably became a major

---

[24] Quoted in W. Link, *Die amerikanischer Stabilisierungspolitik*, p. 56.

[25] See C.S. Maier, 'The factory as society. Ideologies of industrial management in the twentieth century', in R.J. Bullen et al. (eds), *Ideas into Politics* (London, 1984), pp. 148ff.

attraction. His memoirs had been translated into German in 1923 and became an immediate bestseller, running through more than thirty impressions. In 1928, then, Moellendorff was sent on an extensive tour by IG Farben's C. Bosch to 'examine the transferability of experiences to Germany'. Moellendorff, of course, was no newcomer to the field and his report, 'Elementary comparisons of the national economies of the United States, Germany, Britain, France, Italy', contained many telling observations on American business attitudes.[26] Moellendorff was particularly impressed by one point, 'that both the American and the Russian doctrine of salvation are focusing on the idea that a modern national economy should be more concerned with the "poor" than with the "wealthy" consumers'.[27] In line with this postulate and by comparison with Europe, he added, the purchasing power of the small household had indeed seen 'a marked improvement' in relation to the rich.

This was precisely what Ford had been agitating for when he outlined his vision of a mass production and mass consumption society. It was also the point of comparison between Germany and the USA for M.J. Bonn, a respected liberal economist. According to him, American capitalism was moved by motives different from those that usually struck the foreign visitor admiring rationalized and standardized mass production:

> Ford's significance does not lie in [his] assembly-line and a well-thought-out division of labour which the grown-up German children who visit America for the first time see as the *raison d'etre* of American life. Rather it lies in the sober fact which is propagated under the slogan of 'social service' and hence somewhat removed from rational analysis that American entrepreneurs like Ford know that the masses will only tolerate the accumulation of great wealth in the hands of a few, if they themselves derive a corresponding advantage from it. In a wealthy country like America one permits the entrepreneur to earn as much as he likes, provided that those through whom he makes his money also benefit from it.[28]

Bonn furthermore drew attention to the political aspects of Fordist ideology. The pragmatic American businessman, he added, appreciated

---

[26] W. von Moellendorff, *Volkswirtschaftliche Elementarvergleiche zwischen Vereinigten Staaten von Amerika, Deutschland, Grossbritannien, Frankreich, Italien* (Berlin, 1930).

[27] Ibid., Vol. I, p. 6.

[28] M.J. Bonn, *Das Schicksal des deutschen Kapitalismus* (Berlin, 1930), pp. 46ff.

that an economy which was not geared to the 'little man' would, in the age of universal suffrage and democracy, be threatened by the political power of the non-capitalist. Although detailed research is still lacking, it may be assumed that these were also the points which the leading men of Siemens and Bosch would have accepted. These companies were interested not only in Fordist standardization and an expansion of the mass market in consumer goods, but also in the wider societal ramifications of increased productivity. They were, interestingly enough, also among those industrialists who in the mid-1920s were preapred to enter into closer co-operation with the Americans and with the unions. On the other hand, it was precisely these implications which the conservatives in coal and steel found deeply suspect. They recognized and probably respected the industrial potential of the USA. But given their general attitude towards a 'well-ordered' and stratified society and in favour of strict leadership from the top, it was only consistent that they rejected the democratic features of American society and its 'inferior' mass culture,[29] just as they were vigorously opposed to an extension of workers' rights at home.

Thus, if his memoirs provide any guide, Zangen, soon to be one of the most powerful men in the Ruhr, returned from his visit with a very negative image of the United States and, it may be surmised, conversely with his belief in the superiority of German culture and industrial organization fortified.[30] It is against the background of such attitudes that a filtering process set in which Hartmann did not think possible: conservative in their social and political outlook, but nevertheless fascinated by the technical possibilities of the machine age, many German industrialists began to standardize and rationalize their enterprises along American lines.[31] They invested large sums in the latest machinery. Yet they determinedly tried to cut out the rest of Fordist ideology. Instead, while rationalization threw workers on to the streets, they expanded their elaborate system of cartels and syndicates to secure a strictly regulated production of high-priced goods which fewer and fewer could afford to buy. Bonn was particularly scathing about what they were doing:

> The authoritarian German capitalism, and heavy industry in particular, has never allowed others to share in its earnings. Obsessed by technically perfectly correct organizational ideas, it has tried to achieve the

---

[29] General images: P. Berg, *Deutschland und Amerika, 1918–1929* (Lübeck, 1963).
[30] W. Zangun, *Aus meinem Leben* (not published).
[31] Weisbrod, *Schwerindustrie*, pp. 52ff.

removal of all technically dispensable intermediate links. As a result not only the ranks of those who wish to earn a share have been thinned, but also the number of those has been reduced who in their hearts take a benevolent interest in the continuation (of capitalism). Capitalism in America is certainly not superior ethically to that of Germany. It is merely much more clever economically.[32]

The consequence of this difference in approach became apparent during the great slump in the 1930s, as a comparison between the crisis management of the New Deal and the solution found in Germany will show. More specifically, with the onset of the depression, if not before, the conservatives in German industry began to reassert themselves over the 'liberals'. The brief spell of tentative collaboration with the unions and of experimentation with Fordist mass consumption ideas came to an end, and so did, characteristically, parliamentary democracy. The 'authoritarian German capitalism' had never been compatible with the democratic constitutionalism of the Weimar Republic and it was only logical that the authoritarians in industry should start to work for its replacement by an illiberal and anti-democratic regime.[33] Reich Chancellor F. von Papen's 'New State' had all the essentials of such a regime, but it lacked a popular base. The Nazi movement was able to provide this base and when, in the winter of 1932–3, a 'Hitler solution' appeared to offer the only way out of the crisis, the now dominant sections of German industry were, if nothing else, prepared to give it a chance. Nor, once Hitler had been installed, did they find the cornerstones of his policies objectionable: his proscription of the working-class movement, his assertive nationalism *vis-à-vis* the victors of the First World War, his rearmament programme which, while permitting an expansion of rationalized production, shifted manufacture away from consumer goods.

Thenceforth Germany went it alone. Instead of increasing international co-operation the Nazi dictatorship set out to realize older ambitions to establish a formal continental empire through military conquest; instead of promoting the reconstruction of the multilateral world trading system, it severed its former links further and began to

[32] Bonn, *Das Schicksal*, p. 47.

[33] On German industry in the 1930s see Abraham, *Collapse of Weimar Republic*; R. Neebe, *Grossindustrie, Staat und NSDAP, 1930–1933* (Göttingen, 1981); H.A. Turner, *German Big Business and the Rise of Hitler* (Oxford, 1985); J. Gillingham, *Industry and Politics in the Third Reich* (London, 1985); A. Schweitzer, *Big Business in the Third Reich* (Bloomington, 1964).

build up an autarkic economic block.³⁴ Accordingly the filters against American cultural influences were also strengthened in the 1930s. Fordism, as it has been defined above, was replaced by a 'German model' of factory organization in which the harsh *Führer* principle reigned supreme and work 'incentives' were provided by a ubiquitous secret police.³⁵ Consumer goods were given a lower priority, behind the production of tanks and fighter-planes. The broader assumptions about economy and society underlying the cultural exports from the USA were anathema in the 1930s, as the Nazis moved to create a racist *Volksgemeinschaft*. Increasingly, the two countries moved into opposite power-political and ideological camps, until Germany finally made a military bid to escape the hegemonic influence of the 'Anglo-Americans', as they were then called.

The meaning of the Second World War may be examined from a variety of angles. Clearly it was about military and political power and about ideology. But it was also about the life or death of two industrial systems which had meanwhile been developing in different directions. The USA, more than Britain, represented the main antagonist of the Second World War, whose imports, whether material, political or cultural, were to be banned from Germany. If the Nazis had won the war, an economic system would have emerged in occupied Europe which, though still capitalist, would have been guided by principles other than those operating in the USA.³⁶ This also applies to factory organization and production. However, the Third Reich was defeated and with defeat there reappeared the question of American hegemony over Germany, enhanced not only by America's transformation into a superpower, but also by her influence as the major Western power in occupied and divided Germany.

The impact of the direct presence of the USA in postwar Germany was to be felt in all spheres of life and has been analysed respectively.³⁷ However, we cannot concern ourselves here with the American role in the reconstitution of political parties, the educational system or the press. Rather we must come back to the question of cultural exports which Hartmann posed for the postwar period when he researched his

---

³⁴ See D. Petzina, *Autarkiepolitik im Dritten Reich* (Stuttgart, 1968); V.R. Berghahn, *The Americanization of West German Industry, 1945–1973* (New York, 1986), pp. 24ff.; R. Opitz (ed.), *Europastrategien des deutschen Kapitals* (Köln, 1976).

³⁵ T. Mason, *Arbeiterklasse and Volksgemeinschaft* (Opladen, 1975).

³⁶ Berghahn, *Americanization of West German Industry*, pp. 13ff.

³⁷ See, e.g., J. Gimbel, *The American Occupation of Germany* (Stanford, 1968).

Amerikanische Firmen in Deutschland. There can be little doubt that the US authorities were keen to get German industry to adopt its model of industrial production and organization. This ambition was quite unambiguously formulated by R.M. Bissell Jr, the second-in-command in the Marshall Plan Administration in Europe. America, he advised, should exploit

> to the full the example of its own accomplishments and their powerful appeal to the Europeans (and others) among all groups. Coca-Cola and Hollywood movies may be regarded as two products of a shallow and crude civilization. But American machinery, American labor relations and American management and engineering are everywhere respected. The hope is that a few European unions and entrepreneurs can be induced to try out the philosophy of higher productivity, higher wages and higher profits from the lower prices of lower unit costs. If they do, if restrictionism can be overcome in merely a few places, the pattern may spread. The forces making for such changes are so powerful that, with outside help and encouragement, they may become decisive. It will not require enormous sums of money (even of European capital) to achieve vaster increases in production. But it will require a profound shift in social attitudes, attuning them to the mid-twentieth century.[38]

The issues of technological change and the export of industrial culture broached in this chapter could hardly have been put more succinctly than they are in this quotation. And Bissell was right: compared with the industries of war-torn Europe, America was well ahead. Rationalization and standardization had made further strides. Thus the Standardization Committee of German Industry reported in 1953: 'Although the Americans produce 70 per cent of all automobiles in the world, they have no more than sixty different models. Europe by contrast, which has a good quarter of world production, makes well over one hundred models.'[39] The Committee added 'that a reduction of the close to 20,000 piston ring types to 2,000 would set free capital assets of 1.2 million marks in one single factory alone'.

The Bissell quote also shows that wider considerations than these were at stake. As the *Industriekurier*, a paper close to West German industry, put it on 19 August 1952, many American businessmen thought of European industry as being 'quaint'. Europe's managers had more to learn from American industry than the latest machine-manufacturing techniques. No one was more convinced of this than

---

[38] Quoted in Berghahn, *Americanization of West German Industry*, pp. 247ff.
[39] Quoted in *Der Spiegel*, 5 August 1953.

Bissell's superior, Paul Hoffman, a former president of the Studebaker Corporation and then Marshall Plan Administrator for Europe. If he began to promote visits to the USA, he did so in the hope that Europe's industrialists would take back with them not merely American technology, but also managerial practices and organizational methods as well as the larger ideological assumptions behind what they saw across the Atlantic. It would be a most illuminating exercise to retrieve and evaluate the reports which different groups of visitors presumably compiled after their return. Another method of diffusion was through the 'Training-Within-Industry' (TWI) programme designed for those who were unable to go to the USA. This was a programme of courses, first introduced in the American zone of occupation in Germany, by which German industrialists would learn about new methods of personnel management and leadership among middle management. In short, by the early 1950s a major effort was being made by the USA to export both its technologies and its ideas about industrial organization, the principles of Anti-Trust included.[40]

Looking at the German response to these initiatives, we soon discover divisions within industry which are already familiar to us from the interwar period. There were still the conservative 'nationalists' who believed that German industry had little to learn from the US model. They were opposed to American ideas about decartellization and oligopolistic competition, suspecting L. Erhard, the neo-liberal Minister of Economics, of wishing to import 'an American seedling' (*ein amerikanisches Pflänzchen*).[41] They advocated instead the reintroduction of the cartels and syndicates of the pre-1945 period, which had been banned by the occupying powers. In the field of industrial relations they continued to uphold the traditional authoritarianism whose strength was analysed by Hartmann in his *Authority and Organization in German Management*. Nor did they have much time for consensus politics and liberal-corporatist co-operation.[42] It was only consistent with their conservatism that they also disliked American ideas on factory organization and management, which had meanwhile seen a further evolution beyond Fordism and now emphasized team spirit and psychological motivation as a complement of material incentives.[43]

---

[40] Details in Berghahn, *Americanization*, pp. 155ff.
[41] *Frankfurter Allgemeine Zeitung*, 17 November 1953.
[42] See Berghahn, *Americanization*, pp. 182ff.
[43] See Maier, 'Factory as society', pp. 151ff.; Hartmann, *Authority and Organization in German Management*.

There were other German industrialists who, like their predecessors in the interwar period, were open to foreign ideas. They tended to work in branches which had old ties with the world market and, in particular, with American business. They appreciated early on that the by now massive hegemonic weight of the USA would decisively shape the structures and maxims by which the postwar world economy would be run. They were willing to integrate German industry into these structures so as to be able to operate successfully and smoothly under the *Pax Americana*. A man who represented this type of businessman very well was Ludwig Vaubel, soon to become the director general of *Vereinigte Glanzstoff*, the chemicals trust at Wuppertal.[44] In 1950 he had been the first German to complete the Advanced Management Course at the Harvard Business School. His experiences there convinced him that West German industry was lagging far behind that of the USA. On his return he founded the 'Wuppertal Circle', which organized seminars modelled along American lines and designed to help his fellow-industrialists to catch up. O.A. Friedrich, a key manager of the Phoenix Tyre Company at Hamburg-Harburg, provides another example.[45] He could draw on his work experience with Firestone during the 1920s and 1930s and on his discussions with his brother Carl Joachim, the prominent Harvard political scientist. Accordingly, Friedrich was not only interested in American methods of rubber production and tyre moulding, but also began to experiment at Harburg with concepts of employee consultation and participation which were apparently inspired by American-style human relations.

Most importantly of all, 'Americanizers' like Vaubel and Friedrich were no longer in a small minority when they propagated their ideas in the late 1940s and early 1950s. The balance of power within industry had begun to shift, and this time the overall constellation was such that the shift became irreversible. Although the conservatives continued to dominate the *Bundesverband der Deutschen Industrie* (BDI) for some time, the Vaubels and Friedrichs found support in another peak association, the *Bundersvereinigung deutscher Arbeitgeberverbande* (BdA), the national employers' federation. *Der Arbeitgeber*, the federation's journal, openly propounded the adoption of American management methods. Meanwhile, BdA representatives travelled to

---

[44] L. Vaubel, *Zusammenbruch und Wiederaufbau* (Munich, 1985).

[45] V.R. Berghahn and P.J. Friedrich, *Industrie und Politik. Otto A. Friedrich und seine Zeit* (forthcoming).

various parts of West Germany to give talks on what might be learned from the USA.[46]

The long-term effects of this kind of propaganda are not too difficult to discern. The early postwar period saw not only the importation of American machinery and technology, paid for in part by the Marshall Plan, but also the gradual transfer and diffusion of managerial practices and concomitant ideologies which underlay the material exports. The Weimar experience may have shown that it was not completely impossible to take over modern technologies and to filter out the cultural assumptions that tend to go with them. But the costs of the 'anti-American' policies adopted by the conservatives in German industry should also have become clear: Germany's industrial system was pushed in a direction which differed markedly from that of the USA and which contributed to making a contest between the two systems, such as occurred in the early 1940s, ultimately inevitable. The outcome of the Second World War weakened the strong position of the conservatives. There emerged from the defeat a different balance within German industry which facilitated the resumption of American 'cultural imports' on an even larger scale than in the interwar period.

To be sure, the adaptation to the American model was by no means total. What emerged this time was a peculiar mix between indigenous traditions and practices and imports from the hegemonic power across the Atlantic. The same consideration would apply to the cases of France, Britain, Italy or, for that matter, Japan. Whatever future research may tell us about the responses of these countries to the American industrial impact, there can be little doubt that Germany, having dithered in the 1920s and rebelled in the 1930s, proved particularly receptive after 1945 not merely to the importation of US technology, but also to the less tangible elements which the hegemonic power across the Atlantic offered for export from its own industrial culture. This conclusion also opens up new avenues for comparisons between a number of European nations. But it should be clear from the approach in this chapter that these comparisons are quite different in kind from those which preoccupied business historians a generation ago.

---

[46] See Berghahn, *Americanization*, pp. 234ff.

9
Technological Diffusion: the Viewpoint of Economic Theory

Paul L. Stoneman

Introduction

Technological diffusion is the process by which innovations (be they new products, new processes or new management methods) spread within and across economies. Some understanding of the process of technological diffusion is essential if we are to gain any insight into processes of economic growth and development, for, whatever the emphasis has been in the past in research and public policy, it is the application of innovations (diffusion) rather than the generation of innovations (invention or research and development (R & D)) that leads to the realization of benefits from technological advance.

My intention in this chapter, which is my third attempt at surveying this subject,[1] is to consider the different approaches in the economics literature that have been pursued in order to rationalize certain 'stylized facts' that describe the diffusion process. The two stylized facts most commonly noted are: (a) the spread of the use and/or

---

An earlier version of this chapter was presented at the Conference on Innovation Diffusion, Venice, 18–22 March 1986, and published in *Ricerche Economiche*, 4 (1986), pp. 585–606.

[1] See P. Stoneman, *The Economic Analysis of Technological Change* (Oxford, 1983), and 'Theoretical approaches to the analysis of the diffusion of new technology', in S. Macdonald et al. (eds), *The Trouble with Technology* (London, 1983).

ownership of a new technology is a time-intensive process;[2] (b) in many cases, plotting usage or ownership of a new technology against time yields an S-shaped (or sigmoid or ogive) curve.[3]

The economics literature on diffusion has grown apace over the past twenty years. To make my task manageable I am restricting myself to the theoretical underpinnings of this literature. Much of the published work represents implicit or explicit applications of theory to particular case studies. I do not wish to belittle this work as it is most important that our theories should be tested. Here, however, I am more concerned with underlying rationales than observations of outcomes. I will be restricting myself to the economics literature, although other subject areas have interesting points to make on diffusion. We might, however, note that parallels between the work in sociology[4] and geography,[5] and the work in economics are surprisingly frequent.

The spread of a new technology occurs in a number of dimensions. The potential buyers of a technology can be corporations, public institutions (which two I generally class together as firms) or households. The literature has tended to observe and model the diffusion process by considering that intra-firm or intra-household diffusion (i.e. studies of extent of use by individual actors) is a separate process from inter-firm or inter-household diffusion (i.e. studies of the proportion of the population using the technology at *any* positive level). I will continue to follow this useful convention. I will also, for reasons of length and practicality, limit myself to diffusion within a single economy, i.e. I will not discuss the international transfer of technology, although my impression is that the theoretical tools I discuss can just as effectively be applied to that issue.

The structure of the rest of this chapter is conditioned by my own views as to what is and what is not important in this area. The first preconception reflected in the structure is that the observed stylized facts, like all such observations in economics, cannot be the result of solely demand-side phenomena. Thus although the first step is to consider demand based models, I complement this by a study of supply and supply–demand interation. Then I consider the link

---

[2] See, for example, E. Mansfield, *Industrial Research and Technological Innovation* (New York, 1968).

[3] See, for example, Z. Griliches, 'Hybrid corn: an exploration in the economics of technological change', *Econometrica*, 25 (1957), pp. 501–22; S. Davies, *The Diffusion of Process Innovations* (Cambridge, 1979).

[4] See E.M. Rogers, *Diffusion of Innovations* (New York, 1983).

[5] See L.A. Brown, *Innovation Diffusion: a New Perspective* (London, 1981).

between diffusion and R & D. Next, because it is often considered that government could and/or should intervene in the technological diffusion process I consider policy issues. Finally I draw some personal views on the most productive avenues, for future research and provide a short conclusion.

Demand-based models

I often find when discussing technology-related issues that a mention of Schumpeter somewhere near the beginning enables one to catch the attention (if not the sympathy) of the audience. It is appropriate therefore that I should want to start with Schumpeter's views on diffusion. In Schumpeter[6] the diffusion process of major innovations is the driving force behind the trade cycle (the long-term Kondratief cycle), but the forces driving the diffusion process *per se* are not made particularly explicit. The conception is that an entrepreneur innovates and the attractiveness of attaining a similarly increased profit and the pressures on the costs of old technologies in a new regime encourage others to imitate, this imitation representing a diffusion process.

One could attempt to make the underlying theoretical approach more explicit and rigorous, but I find this is somewhat unnecessary. The reason is that, surprisingly perhaps, I consider that some of the most recent and most explicitly mathematical work on diffusion is modelling precisely the process that Schumpeter was describing. Although beginning with this work rather overturns the historical flow of the survey, I find there is some advantages in starting here. The view to which I am referring can be called game theoretic or strategic, but I just prefer to associate it with Reinganum.[7]

The assumption is that there is an n firm industry in which all firms are the same and all have perfect information. There is a new cost-reducing, capital-embodied process innovation available, the cost of adopting which falls over time (at least until some distant date). It is then shown that under certain conditions an equilibrium exists in which the firms in the industry adopt at different dates and thus a diffusion path may exist. The actual proofs in the papers are not simple

---

[6] J.A. Schumpeter, *The Theory of Economic Development* (Cambridge, Mass., 1984).

[7] J. Reinganum, 'On the diffusion of new technology: a game theoretic approach', *Review of Economic Studies*, 48 (1981), pp. 395–405; 'Market structure and the diffusion of new technology', *Bell Journal of Economics*, 12 (1981), pp. 618–24; 'Technology adoption under imperfect information', *Bell Journal of Economics*, 14 (1983), pp. 57–69.

and the presentation is very mathematical. I will here, therefore, present a more conceptual version of the argument, without, I hope, too much cost to the original.

Let the present value of firm i's profits when m other firms are using the new technology be $\pi_1(m,n)$, if it is also a user and $\pi_0(m,n)$ if it is a non-user. Both $\pi_1(m,n)$ and $\pi_0(m,n)$ will decline with m (and n), for as use of the technology extends firms will increase their output and prices will fall. Of course $\pi_1(m,n) > \pi_0(m,n)$. Let the costs of adopting the technology in period t be C_t. A firm behaving myopically (on which I say more below) will adopt the new technology[8] in time t if

$$\pi_1(m,n) - \pi_0(m,n) \geq C_t \tag{1}$$

i.e. if the profit gain from use is greater than the cost of adoption. On the assumption that the profit gain declines with m (i.e. that the further the firm is down the adoption queue the lower is its profit gain), and also that when m equals n the profit gain is less than C_t, then eqn (1) as an equality will yield $m < n$, and diffusion will not be instantaneous. There is of course no way of identifying which of the identical firms is a user and which is a non-user. To produce a diffusion path, C_t is assumed to fall over time and as it does so usage increases.

I find the Reinganum model particularly useful as a starting point,[9] for (a) it assumes that all firms are the same and have perfect information, which as will be seen is equivalent to removing the basis for diffusion in a number of the models discussed below, yet it still produces a time-intensive diffusion process; and (b) the structure of the model illustrates the basic form of many diffusion models – a decision theoretic modelling of choice behaviour to determine ownership or use at a point in time allied with some changes in underlying conditions over time that generate changes in ownership.

To proceed from here I pursue this structure by next looking at models where potential buyers differ from one another, and then turn to the information-based approaches.

Probit models

In Reinganum's framework firms were the same, information was

---

[8] Intra-firm diffusion is assumed instantaneous, *all* production being immediately transferred to the new process.

[9] One might immediately observe, however, that it is probably of little use for analysing consumer good innovations where one expects strategic behaviour among potential buyers to have very little role to play.

perfect but benefits from acquisition declined as usage extended. In the class of models we consider here we maintain the assumption that information is perfect but we assume that acquirers differ from each other. An acquirer's benefits do not decline as usage extends. The diffusion model is then constructed as follows.[10] Let the potential adopters of a technology differ according to some yet to be specified characteristic, z, that is distributed across the population as $f(z)$ with a cumulative distribution $F(z)$ (see figure 1). Allow that in time t, a potential adopter, i, will be a user of the technology if his characteristic level $z_i \geq \bar{z}_t$, some critical level of the characteristic. Then the proportion of the population who have adopted by time t is given as $1 - F(\bar{z}_t)$. This is shown in figure 1 as the shaded area. As time proceeds either $f(z)$ is assumed to shift or $\bar{z}_t$ changes (falls) and as such events occur the proportion of users and the number of users change, thus tracing out the diffusion path.

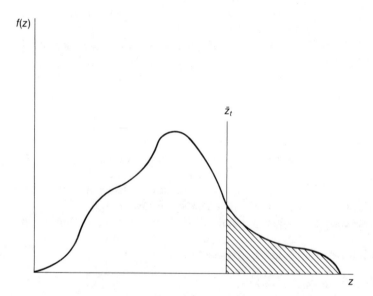

Figure 1 The distribution of firm characteristics

[10] See P.A. David, *A Contribution to the Theory of Diffusion*, Stanford Center for Research in Economic Growth, Memorandum no. 71 (Stanford, Cal., 1969).

Obviously this framework needs a great deal of meat added to it before it is operational. However, being very general it can be used as a general form for analysing many different approaches to diffusion by specifying the appropriate characteristic. I will be concentrating below on the use of the model to analyse the diffusion of a process innovation across firms, so I will mention other uses here. One might, for example, think of a consumer good innovation whereby the relevant characteristic might be socio-economic group or, as used by Bonus,[11] household income. Then ownership extends down the income distribution. We might think of technologies whereby spatial characteristics are important and the distribution reflects this with diffusion following some spatial pattern. Whatever the application the principle is the same.

What is clear though is that we cannot get far until we define the characteristic and the determination of $\bar{z}$. Two of the most successful applications of this approach have considered firms buying process innovations and have isolated firm size as the relevant characteristic, although David refers to applications, considering characteristics such as entrepreneurial attitudes and vintage or age-of-capital effects (vintage models).[12] These applications to process innovations centre upon profit as the criterion for the determination of $\bar{z}$. Thus, in David, the technology is assumed to exhibit increasing returns to scale and at any time there is a certain size of firm above which it is profitable to adopt, below which it is not. This then defines a critical firm size. Davies's model, although not in the profit-maximizing mould, is very similar.

To illustrate more fully the operation of the model I will not be specific about the characteristics but I will keep the profit motive as central. To proceed we make two further assumptions: (1) that when technology is acquired, the firm transfers its whole production to the new process (there is no intra-firm diffusion process); and (b) that the technology can be acquired by purchase of one unit of a new capital good whatever the level of use by the acquirer.[13] Allow that a firm of characteristic level z can by acquisition in time t obtain an increase in its profit flow of $h(z)$ in perpetuity, yielding a present value gain of $h(z)/r$ where r is the discount rate. Also allow that in time t the cost of

---

[11] H. Bonus, 'Quasi-engel curves, diffusion and the ownership of major consumer durables', *Journal of Political Economy*, 81 (1973), pp. 655–77.

[12] David, *Contribution to Theory of Diffusion*; Davies, *Diffusion of Process Innovations*.

[13] By handwaving and, say, the treatment of firms as plants, the severity of these restrictions can be relaxed.

acquisition is p_t, with the expected cost of acquisition in the following period being p^e_{t+1}. A profit-maximizing firm will acquire in time t if (a) it is profitable to do so,

$$h(z)/r \geq p_t \tag{2}$$

and (b) it is not profitable to wait until time $t+1$, i.e.

$$-p^e_{t+1} + (1+r)p_t \leq h(z) \tag{3}$$

One may note that if the buyer is myopic ($p^e_{t+1} = p_t$), then eqn (3) collapses to eqn (2) and the profitability condition is sufficient to determine use. If buyers have perfect foresight and p^e_{t+1} P_{t+1} and $p_{t+1} < p_t$, then satisfying eqn (3) satisfies eqn (2) and eqn (3) is going to determine use.

Define z_t as the characteristic level of the marginal adopter in time t. Then, under myopia,

$$h(z_t) = rp_t. \tag{4}$$

If the potential population of adopters is of size N, then the number of users in time t, M_t, is

$$M_t = N(1 - F(z_t)) \tag{5}$$

$$\therefore \quad \frac{N - M_t}{N} = F(z_t)$$

$$\therefore \quad z_t = F^{-1}\left(\frac{N - M_t}{N}\right)$$

$$\therefore \quad p_t = \frac{1}{r} h[F^{-1}\left(\frac{N - M_t}{N}\right)] \tag{6}$$

Given that each firm only buys one unit of the new capital good the stock of the capital good acquired by time t, $X_t = M_t$. Thus

$$p_t = \frac{1}{r} h[F^{-1}\left(\frac{N - X_t}{N}\right)] \equiv \frac{1}{r} g(X_t) \tag{7}$$

Equation (7) is an inverse stock demand function that relates the stock of the new capital good demanded to the current price of that capital good. If one assumes that buyers are not myopic one can define equivalently to eqn (7) the condition (8) as the inverse stock demand function under perfect foresight.

$$p_t = \frac{h}{1+r}[F^{-1}(\frac{N-X_t}{N})] + p^e_{t+1}$$
$$= \frac{1}{1+r}g(X_t) + \frac{p^e_{t+1}}{1+r} \qquad (8)$$

Which may also be written as

$$(p_t - p^e_{t+1}) + rp_t = g(X_t) \qquad (9)$$

From eqn (7) it is easy to see how a diffusion path results. Holding r constant, totally differentiating eqn (7), and using the dot convention for a time derivative

$$\dot{p}_t = \frac{1}{r}g_x \dot{X} \qquad (10)$$

and $\ddot{p}_t = \frac{1}{r} \dot{X}g_{xx} + g_x \ddot{X})$ \qquad (11)

From eqn (10) usage extends as price falls, given $g_x < 0$. From eqn (11), $\ddot{X}$, which tell us whether the diffusion curve is sigmoid, depends on $\ddot{p}$, g_x, and $\dot{X}$ and g_{xx}. Thus the nature of the diffusion path depends on movement in prices over time and the $g(X)$ function. This in turn will depend upon the $h(z)$ function, i.e. how benefits are related to characteristics, and the $f(z)$ function, i.e. how the characteristics and thus the benefits are distributed across the population. As these functions vary so the diffusion path will be changed. By manipulating these distributions, Davies, for example, illustrates how the diffusion pattern may be related to the concentration of firm sizes.

A variety of models with demand structures on these lines have been constructed.[14] Two of these papers, David and Olsen (1984) and Ireland and Stoneman (1986), are addressed particularly to an issue

---

[14] P.A. David and T. Olsen, 'Anticipated automation: a rational expectations model of technological diffusion', Center for Economic Policy Research, Publ. no. 24 (Stanford, Cal., 1984); N.J. Ireland and P. Stoneman, 'Order effects, perfect foresight and intertemporal price discrimination', *Recherches Economiques de Louvain*, 51 (1985), pp. 7–20; Ireland and Stoneman, 'Technological diffusion, expectations and welfare', *Oxford Economic Papers*, 38 (1986), pp. 283–304; Stoneman and David, 'Adoption subsidies vs. information provision as instruments of technology policy', *Economic Journal* (1986), RES/AUTE Conference supplement; Stoneman and Ireland, 'The role of supply factors in the diffusion of new process technology', *Economic Journal* (1983), supplement, pp. 65–77; Stoneman and Ireland, 'Innovation and diffusion – the implications of an integrated approach', paper presented at the Summer Workshop on Technological Change, Warwick University (1984).

raised by Rosenberg that we can begin to raise in this context.[15] Rosenberg argued that the literature on diffusion took insufficient account of expectations in the diffusion process. He hypothesized that a technology that is expected to improve over time may experience slower diffusion than one that is not. From eqn (9), given $g_x < O$, we can see that for a given p_t, the lower is p_{t+1}^e, the smaller will be X_t, the level of use, i.e. firms will prefer to wait before acquisition. It may be profitable to acquire now, but it is more profitable to acquire tomorrow. Rosenberg's hypothesis thus carries some support when expectations on prices are being discussed. Ireland and Stoneman (1986) consider the issue further. Expectations may refer to technological obsolescence, and they show that in certain circumstances the discount rate r can be interpreted as including an element reflecting the risk of obsolescence, so a heightened expectation of the new technology being supplanted can be modelled by a higher value of r. From eqns (7) or (8) one can see that a higher r would be associated with less use, *ceteris paribus*. Balcer and Lippman consider the case of technological expectations more explicitly with similar results.[16]

These probit models suggest therefore that diffusion is the result of movements down some benefit distribution. The speed of diffusion and shape of the diffusion curve depend on the shape of the benefit distribution and the rate at which movement down the distribution occurs. For the moment the rate of movement is being treated as exogenous; we will have occasion below to make it endogenous. The model's preductions as to the determinants of the diffusion path for empirical study[17] do depend on the isolation of the crucial characteristics.

The model presented here is free of strategic or information complications so that the basis of the approach could be isolated. It is worth noting, however, that David and Olsen (1984) do incorporate some strategic elements, and Stoneman and David (1986) graft information issues on to the basic model. As we shall see in the next part, information-based models often themselves revert to probit type models to generate inter-firm diffusion paths.

---

[15] N. Rosenberg, 'On technological expectations', *Economic Journal*, 86 (1976), pp. 523–35.

[16] Y. Balcer and S.A. Lippman, 'Technological expectations and adoption of improved technology', *Journal of Economic Theory*, 34 (1984), pp. 292–318.

[17] Some empirical studies have been done by P.A. David, 'The mechanization of reaping the ante-bellum Midwest', in H. Rosovsky (ed.), *Industrialization in Two Systems* (New York, 1966); Davies, *Diffusion of Process Innovations*; G. von Tunzleman, *Steam Power and British Industrialisation to 1860* (Oxford, 1978).

Risk, uncertainty, information and learning

Thus far I have ignored one particular characteristic of the diffusion process that should not be ignored and is perhaps of overriding importance. As diffusion concerns something that is new it is probably taking place in an environment where information is imperfect and as a result involves uncertainty and risks.

If the world is one involving uncertainty then the modelling of the decision process must treat decision-making under uncertainty. Thus the several approaches considered in this section have at their core such theories. The diffusion models all need to ally some learning or information acquisition process with their decision-making frameworks. The sources of information can be divided into two classes, external and internal (to the decision-making unit), which gives a possible dichotomy in the classification of diffusion models. A further possible classification in this area is that between inter-firm and intra-firm diffusion models. Some models (inter-firm) consider just whether the potential buyer acquires the new technology or not, others (intra-firm) consider the level of use of the technology by a buyer.

The starting point is epidemic models. In their simplest inter-firm form it is assumed: (a) that upon learning of the existence of the new technology a potential acquirer will adopt the technology (the decision theoretic framework); and (b) that information on existence is spread by personal contact, in that whenever a potential acquirer who is a non-user meets a user, then he obtains knowledge of existence (this information-spreading mechanism obviously leads to the label 'epidemic').

This very simple combination will generate a time path of diffusion that is logistic, which functional form is often used to approximate the diffusion curve, and the speed of diffusion is related to the frequency of inter-personal contact. The simplicity of the model has led to a number of criticisms[18] on the grounds that the potential adopters are considered homogeneous, no allowance is made for technology improving over time, the information-spreading mechanism is too simple and takes no account of other information sources (e.g. advertising) and the decision theoretic framework does not really capture the essence of decision-making under uncertainty (to say the least). Even

---

[18] See, for example, Davies, *Diffusion of Process Innovations*.

so, the model has a distinguished background in the subject.[19]

One variation on the basic epidemic model is developed by Lekvall and Wahlbin.[20] They consider that the communication mechanism discussed above (contact with users) should be supplemented with information coming from outside the set of adopters (through, for example, advertising and other promotional activities). They show that this will produce a diffusion curve that is no longer logistic, the external influences modifying its shape. We might note that both Gould and Glaister have looked further at the role of advertising in diffusion models of this kind.[21]

Despite these modifications to the information mechanism, such models as these are particularly weak on the decision theoretic side. This leads to what is probably the most widely known and widely used diffusion model, that of Mansfield.[22] This model comes in two forms, the inter-firm and intra-firm models, which are largely distinguished by whether the information source is internal or external to the firm. In the inter-firm model, the decision theoretic framework is represented by a reduced form hypothesis. It is argued that at any point in the diffusion process the number of users acquiring the technology at that moment is related to the risk attached to acquisition, the expected profitability of acquisition and the number of potential adopters. The latter two are assumed to be invariant with respect to time, but risk is assumed to reduce as the number of users increases, thereby driving the diffusion process. The reductions in risk are assumed to come about because increased ownership or use increases knowledge and this reduces uncertainty. By the choice of appropriate functional forms the model generates a logistic diffusion curve with the speed of diffusion linearly related to, *inter alia*, profitability.

The intra-firm variant is very similar. The reduced form decision rule relates the firm's extension of use in a period to risk, expected profit and its final level of use. The latter are assumed invariant to time but risk is reduced over time as the firm learns from its own use. Again,

---

[19] See, for example, Griliches, 'Hybrid corn'; G. Pyatt, *Priority Patterns and the Demand for Household Durable Goods* (Cambridge, 1963).

[20] P. Lekvall and C. Wahlbin, 'A study of some assumptions underlying innovation diffusion functions', *Swedish Journal of Economics*, 75 (1973), pp. 362–77.

[21] J.P. Gould, 'Diffusion processes and optimal advertising policy', in E.S. Phelps et al., *Microeconomic Foundations of Employment and Inflation Theory* (New York, 1970), pp. 338–68; S. Glaister, 'Advertising policy and returns to scale in markets where information is passed between individuals', *Economica*, 41 (1974), pp. 139–56.

[22] Mansfield, *Industrial Research*.

by appropriate choice of functional forms, a logistic diffusion curve with diffusion speed linearly related to, for example, profitability is generated.

Despite the undoubted empirical success of the generated estimating equation from these models, the framework has been subject to various criticisms:

1. The information sources are all internal in the sense used by Wahlbin – risk is only reduced as the number of users extends in the inter-firm model or as the firm's usage extends in the intra-firm model; for example, advertising has no role to play.
2. It is difficult to see exactly from what decision theoretic framework the reduced form adoption rule is derived.
3. Technology is assumed to not change over time (see Gold, who describes the model as like filling a bottle).[23]
4. The treatment of risk, uncertainty and information acquisition has been considered very inadequate.[24]

To some extent such criticisms as these have been overcome by recent work that more explicitly details the process of decision-making under uncertainty. These approaches have been applied extensively in the analysis of the adoption of agricultural innovations in developing countries. A recent survey by Feder, Just and Zilberman summarizes much of the theoretical and empirical literature in this area, a fair amount of which can be attributed to Feder and co-workers.[25] It appears that, simultaneously, this group and others were all working in

---

[23] B. Gold, 'Technological diffusion in industry: research needs and shortcomings', *Journal of Industrial Economics*, 24 (1981), pp. 247–69.

[24] See, for example, P. Stoneman, 'Intra firm diffusion, Bayesian learning and profitability', *Economic Journal*, 91 (1981), pp. 375–88.

[25] G. Feder, R.E. Just and D. Silberman, 'Adoption of agricultural innovations in developing countries: a survey', *Economic Development and Cultural Change*, 33 (1985), pp. 255–98; Feder, 'Farm size, risk aversion and the adoption of new technology under certainty', *Oxford Economic Papers*, 32 (1980), pp. 263–83; Feder, 'Adoption of interrelated agricultural innovations: complementarity and impact of risk, scale and credit', *American Journal of Agricultural Economics*, 64 (1982), pp. 94–101; Feder and G.T. O'Mara, 'Farm size and the adoption of green revolution technology', *Economic Development and Cultural Change*, 30 (1981), pp. 59–76; Feder and O'Mara, 'On information and innovation diffusion: a Bayesian approach', *American Journal of Agricultural Economics*, 64 (1982), pp. 141–5; G. Feder and R. Slade, 'The acquisition of information and the adoption of new technology', *American Journal of Agricultural Economics*, 64 (1982), 145–57.

slightly different ways towards the same objective – modelling diffusion under uncertainty.[26]

The work of Stoneman and Lindner et al. is very similar. A firm is considered that has a choice of using new or old technology. Each technology has associated with it a mean and variance of returns, which are known for the old technology but not for the new. The firm makes decisions on the basis of maximizing a utility function defined on the overall mean (profitability) and variance (risk) of its production mix, by the appropriate choice of the proportions in which the two technologies are being used. In the initial period the firm forms a prior estimate of the distribution of returns to the new technology. As time goes on this prior estimate is updated as new information is gathered. The updating is modelled as a Bayesian process. The updated estimates lead to changes in the desired level of use, thus tracing out the diffusion process.

Lindner et al. consider the problem of first use of a new technology and allow that the information sources are external to the firm. Adoption (in terms of first use) will occur when the estimate of the mean return to the new technology is high enough to overcome the risk (variance) attached to it. The Stoneman version is more concerned with the proportion of the firm's output produced on the new technology once initial adoption has occurred. In this model all information is derived from internal sources. It is thus basically a model of intra-firm diffusion. It is shown that at a moment in time the level of use will depend on: (a) the true mean and variance of returns to the new technology; (b) the mean and variance of returns to the old technology; (c) the initial estimates of the mean and variance of returns to the new technology; (d) the firm's attitude to risk; and (e) the correlation of the returns to the new and old technology.

---

[26] R.A. Jensen, 'Adoption and diffusion of an innovation of uncertain profitability', *Journal of Economic Theory*, 27 (1982), pp. 182–93; 'Innovation adoption and diffusion where there are competing innovations', *Journal of Economic Theory*, 29 (1983), pp. 161–71; 'Adoption of an innovation of uncertain profitability with costly information', Dept of Economics working paper 84-8, Ohio State University (1984); 'Innovation adoption with both costless and costly information', Dept of Economics working paper 84-22, Ohio State University (1984); 'Information capacity and innovation adoption', Dept of Economics working paper 84-33, Ohio State University (1984); P. Stoneman, 'The rate of imitation, learning and profitability', *Economics Letters*, 6 (1980), pp. 179–83; Stoneman, 'Intra firm diffusion'; Stoneman, *Economic Analysis of Technological Change*; Stoneman, 'Theoretical approaches to the analysis of diffusion of new technology'; R. Lindner, A. Fischer and P. Pardey, 'The time to adoption', *Economic Letters*, 2 (1979), pp. 187–90.

These models are simply converted to inter-firm diffusion models. Define α_{it} as the proportion of ith firm's output produced on the new technology in time t, and define $\hat{\alpha} > 0$ as the level of α above which the firm is defined as a user and below which it is defined as a non-user. One may then proceed in the manner of probit models. Given the distribution across all firms of α_{it} at a point in time one can define the proportion of firms for which $\alpha_{it} > \hat{\alpha}$, and thus generate the extent of diffusion. As α_{it} changes over time so will the extent of diffusion. Given the determinants of α_{it} above (a to e) one has predictions as to which characteristics are associated with early users, which with late users and which will determine the speed of movement along the diffusion path.

Tonks has criticized the work of Stoneman on the grounds that the model does not allow the firm to buy technology in order to acquire information, that the information is being treated as a non-valued by-product of use.[27] The Tonks model tries to correct this. It is also unique in this area in concentrating on consumer good innovations rather than producer good innovations. However, the important basic principle of the model is that economic agents will, given that information is valuable, actually undertake a search rather than wait for information to arrive.

Jensen has considered this possibility and is one of the few authors to do so.[28] In his early papers, however, he does not. In his 1982 paper he considers a firm that receives external information about a potential innovation. It does not know whether the innovation will be profitable or not, but the information allows it to learn. Adoption is an all or nothing decision. The firm is assumed to use Bayesian updating rules and at each moment can make one of three decisions: (a) acquire, (b) not acquire, or (c) await further information before making a decision. The decision to adopt is considered irreversible and decision-makers are risk neutral, acting to maximize expected returns. The decision problem is shown to be an optimal stopping problem, and the optimal behaviour of the firm is shown to be to adopt when its current belief that the innovation is good is above a minimum reservation level, which in turn is dependent on the cost of adoption, the return to the technology and the discount rate. Setting the decision up in this way leads obviously to an inter-firm diffusion model using probit methods.

---

[27] I. Tonks, 'Advertising, imperfect information and the effect of learning on consumer behaviour', PhD Thesis, University of Warwick (1983); 'The demand for information and the diffusion of a new product', mimeo, University of Exeter (1985).

[28] Jensen, 'Adoption and diffusion of an innovation of uncertain profitability'.

In Jensen's 1983 paper the model is extended to choice between competing innvoations. In both cases it is shown that S-shaped diffusion curves can be predicted. Reinganum explores a very similar model.[29]

In his 1984 papers, Jensen extends the model to allow more than one information message per period, but most interesting are the two papers where the information is treated as costly to the firm, i.e. the firm has to pay for information.[30] Jensen argues that if learning is costly then immediate adoption becomes more likely, delayed adoption becomes less likely, and eventual adoption of a profitable innovation is not certain. This is as one would expect: costly information will only be acquired if the expected gain from having that information is greater than its cost. Some firms may thus never learn about an innovation that is profitable to them for they are unwilling to pay for the information.

This growing literature has a number of useful results. It emphasizes attitudes to risks, prior and updated estimates of risks and returns etc. as important in the diffusion process. It also allows firms to make mistakes, so unprofitable innovations may be adopted leading to a possible eventual reversal of the diffusion process. Even so, there are problems: in particular, as far as I know, this literature ignores the possibility that technology may improve or become cheaper over time and it still tends to treat diffusion as 'filling a bottle'.

Behavioural and evolutionary models

All the frameworks described above, although they might include imperfect information and uncertainty, maintain the long neo-classical economic tradition that economic actors are maximizers. An alternative line may be that firms satisfice or use rule-of-thumb decision-making processes. Perhaps the main proponents of this view today in the technological change literature are Nelson and Winter with their evolutionary models. I have also published some work on this line.[31] The work is not easy to summarize briefly. The main characteristics of the approach are those that will be familiar to students of behavioural

---

[29] Reinganum, 'Technology adoption under imperfect information'.

[30] Jensen, 'Adoption of an innovation of uncertain profitability with costly information'; 'Innovation adoption with both costless and costly information'.

[31] R. Nelson and S.G. Winter, *An Evolutionary Theory of Economic Change* (Cambridge, Mass., 1982); P. Stoneman, *Technological Diffusion and the Computer Revolution* (Cambridge, 1976).

theory, *i.e.* satisficing, local search, problem-orientated decision-making etc. Nelson and Winter discuss diffusion in a framework very similar to the framework used here, i.e. decision processes, information-spreading processes, risk and search etc. One of their main contributions, although not completely absent from the discussions above, is their emphasis on variety. Different innovations in different industries will have different diffusion patterns (agriculture is not like aircraft); public corporations may behave differently from private firms; and the regulatory environment may also affect the diffusion path. We are warned therefore to be wary of over-generalization.

Supply and demand

The modelling frameworks I have been discussing have considered the demand side in the diffusion process. However, although such demand models have in some cases been used alone to explain observed diffusion phenomena, in general any realized diffusion pattern must be the result of not just a demand pattern but of a supply–demand interaction. This requires, if one is fully to understand the diffusion process, that one has some insight into the supply side as well as the demand side, and some knowledge of the interaction between supply and demand. I am going to assume that the typical innovation is produced in one industry and then sold to another industry or to the consumer. It may be for some innovations that the buyer and seller are in the same industry, but I do not discuss that case. I will also assume that the supply industry is a domestic industry. Some technologies may be bought from overseas but I do not consider that case. One may note that for the supplying industry the new technology will be a product innovation, for the buying industry it will be a process innovation.

The modelling of the supply side is designed to generate (a) a supply curve relating for each time period the quantity at each price the industry is willing to supply, (b) a time path for technological improvements and (c) a resolution of any conflict between quantities supplied and demanded. To investigate such issues it is clear that one must model (a) the time structure of the supply industry's costs, (b) the capacity of firms in the supplying industry, (c) the number of suppliers, (d) the price and quantity setting behaviour of firms and (e) the nature of market interactions; stated in this way it soon becomes obvious that such issues represent a major research area that is not only of interest in the study of diffusion. However, having said this, our knowledge of supply industry behaviour especially in the context of diffusion is very limited. To illustrate this point, I have argued that for the supply industry new technology will represent a product innovation. Competition between firms may thus come down to product innovation competition, and the appropriate way to model such a market is a method involving product differentiation. As economists we are limited in our ability to do this at the present time. One possible approach is illustrated in the work of Shaked and Sutton but as far as I know this has not been applied to the diffusion problem as yet.[32] A major research opportunity exists here.

To illustrate the work that has been carried out I start by considering Ireland and Stoneman.[33] In this paper the supply industry is assumed to be an n firm symmetric oligopoly (of which $n = 1$, monopoly, and $n \to \infty$, perfect competition, are the special cases). Firms are assumed to be quantity setters with Cournot conjectures. Firms are assumed to know the demand regime (which is of the probit type as represented in eqns (7) or (9) above, and maximize their present values. The costs of production are assumed to fall exogenously over time. Diffusion proceeds by firms reducing prices over time, which increases use by movements down the reservation price distribution over potential users. In Ireland and Stoneman the main question approached is how the buyer's expectation regime will affect the diffusion path. The results indicate that: (a) for a given number of suppliers diffusion will be faster if buyers have perfect foresight on prices rather than hold myopic expectations; (b) given the expectations regime, the greater the number of suppliers the faster is diffusion; but (c) perfect competition in supply with buyers having perfect foresight yields the same diffusion curve as a monopolist supplier combined with buyers who are myopic.

There are now a number of variants on this model. One variant is to consider that suppliers' costs do not fall exogenously but reduce with accumulated output (learning by doing). An important issue here is whether the learning by doing is firm-specific or industry-wide. The learning by doing variant is explored by David and Olsen, and Stoneman and Ireland.[34] David and Olsen also incorporate a slightly

---

[32] A. Shaked and J. Sutton, 'Relaxing price competition through product differentiation', *Review of Economic Studies*, 49 (1982), pp. 3–13; 'Natural oligopolies', *Econometrica*, 51 (1983), pp. 1469–83.

[33] Ireland and Stoneman, 'Technological diffusion, expectations and welfare'.

[34] David and Olsen, 'Anticipated automation'; Stoneman and Ireland, 'The role of supply factors'.

different demand structure in their model, with some strategic elements in it. Another probit variant with early mover advantages on the demand side is explored by Ireland and Stoneman.[35] I know of work proceeding with a demand side modelled on Reignanum lines, but I do not know of any work adding a supply side to a demand side framework in which decision-making under uncertainty is modelled formally. There is an early paper by Glaister that treats demand as epidemic model based, and Stoneman and David combine a supply side with a mixed epidemic/probit model.[36]

Although this does not represent an exhaustive list of work proceeding on these lines, it suggests that the area is very active. However, there are problems with this work. Not the least of these is that as constructed these models assume that over the diffusion period the number of suppliers is constant. The empirical work of Gort and Klepper shows that this is not a reflection of reality.[37] Some modelling attempts have been made to make the number of suppliers endogenous, but they have not really come to grips with the question: what determines the number of producers of a new product?[38] I do not really think that economics has a complete answer to this. Spence has made some advances in this area, and I have attempted to survey the relevant literature.[39] It seems to me, however, that there is still much to do here. I do in fact return to this issue below.

The work already referred to has other limitations. First it assumes that markets always clear. There is some evidence that firms' prices will not always clear the market and orders or inventories may build up.[40] As far as I know such non-price adjustment mechanisms have not been formally modelled. Secondly, except in the case of Glaister, the possible use by the suppliers of advertising or other forms of non-price competition tends to be ignored. Finally, the models so far discussed tend to abstract from the problems of capacity. A firm selling a new product for which demand is likely to grow and then fall faces a number of problems: (a) should it install capacity to meet peak demand

---

[35] Ireland and Stoneman, 'Order effects'.

[36] Stoneman and David, 'Adoption subsidies vs. information provision'.

[37] M. Gort and S. Klepper, 'Time paths in the diffusion of product innovations', *Economic Journal*, 92 (1982), pp. 630–53.

[38] For example, see Stoneman and Ireland, 'Innovation and diffusion'.

[39] A. Spence, 'Investment, strategy and growth in a new market', *Bell Journal of Economics*, 10 (1979), pp. 1–19; Spence, 'The learning curve and competition', *Bell Journal of Economics*, 12 (1981), pp. 49–70; Stoneman, *Economic Analysis of Technological Change*.

[40] Stoneman, *Technological Diffusion and the Computer Revolution*.

or try to smooth demand and hold reduced capacity; (b) can it raise sufficient funds to provide the capacity required; (c) how should it fund its capacity creation? Metcalfe has been a major contributor to work on diffusion that stresses this capacity problem on the supply side.[41]

In consider the addition of the supply side to be the most important advance in the economics of diffusion of the past ten years. This is not only because it throws new light on the determinants of the diffusion process (supply industry structure, costs, behaviour etc.) but also because it allows us to make two further advances, the first in linking diffusion to R & D, the second in the policy area.

Diffusion and R & D: the interaction

There has been some limited debate in the literature on the impact that R & D by a potential user can have on the diffusion process.[42] The R & D so discussed can be expenditures on adapting technology to a firm's particular circumstances or R & D as search expenditures. In terms of formal modelling this is largely the same as considering search to be costly and I considered it above by reference to the work of Jensen. Here I am after a different link.

When one adds a supply side to the diffusion process, we know that the number of suppliers and their costs and the improvements in technology that they generate are important influences on the diffusion path. However, production costs, improvements in technology and entry to an industry are largely the result of R & D spending. The incentive to do R & D is expected profitability. This profitability is derived by the suppliers from sales during the diffusion process. Thus the diffusion process generates the incentives to R & D and R & D brings forth the lower costs, improved technology etc. that drive the diffusion. At the risk of over-emphasis, the point being made is that R & D (or invention and innovation) and diffusion are not separate processes. They are in fact an integrated process. The integrated nature of this process has not been fully realized in the literature. There is some formal modelling in Stoneman and Ireland, and Metcalfe has

---

[41] J. Metcalfe, 'Impulse and diffusion in the study of technical change', *Futures*, 5 (1981), pp. 347–59; Metcalfe, 'On technological competition', paper presented to Workshop on New Technology, Windsor (1985).

[42] See, for example, D. Mowery, 'Economic theory and government technology policy', *Policy Sciences*, 16 (1983), pp. 27–43.

been approaching issues on a similar line of enquiry.[43] Perhaps of more importance, however, is that it brings into diffusion analysis the huge body of work on R & D, although I have no intention of summarizing it here. This R & D literature is concerned with the generation of new products or product improvements, with new processes or process improvements, with technological competition between firms etc., all of which are factors that underlie the supply side, the role of which in the diffusion process I have already emphasized. My message is that it may be time to think of re-integrating the Schumpeter trilogy.

Public policy

Policies on diffusion have in most economies been the poor relation to R & D policies in overall technology policy strategies. Despite the fact that the impacts of new technology only arise as new technology is diffused and inventions and innovations that are not diffused have no impact on the economy, it is the support of invention and innovation that has taken the lion's share of most technology support programmes. Diffusion policies, where they have been put into effect, tend to be small and generally of two types: information based policies, e.g. the US Agricultural Extension Scheme; or subsidy policies, e.g. the UK government in the late 1960s subsidizing the purchase of digital computers.

As the conception of the diffusion process has broadened to encompass the supply side and the links with R & D, the view that is taken of policy initiatives has changed. These changes are best treated in a list.

(1) I do not think it a misconception of much of the early writing in this area to say that the view commonly held was that 'technological change is good and faster technological change is better' and thus faster diffusion should always be encouraged. There are some hints of dissension from this view.[44] Recent advances, however, have allowed one to be more explicit. The supply–demand models of diffusion allow one to characterize welfare optimal diffusion paths. Welfare optimality

---

[43] Stoneman and Ireland, 'Innovation and diffusion'; Metcalfe, 'On technological competition'.
[44] For example, C.A. Tisdell, *Science and Technology Policy* (London, 1981); R. Nelson, M. Peck and E. Kalachek, *Technology, Economic Growth and Public Policy* (Washington, DC, 1967).

does not always imply the fastest possible take up of technology and thus the maxim that faster is better does not always hold. Obviously, deciding on optimality requires statement of the welfare objective function. In the various papers by Stoneman and Ireland discussed above, welfare is defined as the sum of suppliers' profits and users' profit or utility gain. This could be extended in the case of process innovations to include additions to buyers' consumer surplus.[45] Ireland and Stoneman show in their 1986 paper that the welfare optimal path is generated by either a competitive supplying industry with buyers with perfect foresight or by a monopoly supplying industry with buyers who are myopic. Myopia with a competitive supply industry yields diffusion that is too fast, perfect foresight with monopoly supply yields diffusion that is too slow. (To illustrate that the nature of the optimal path is rather model specific we might note that David and Olsen have a different welfare optimal path because the costs of producing the new technology fall by learning by doing, whereas in Ireland and Stoneman they fall exogenously. To illustrate further the fragility of the result, the Ireland and Stoneman result is dependent on the returns to the new technology being unaffected by the diffusion path.) However, under perfect competition suppliers get no profit, and under monopoly suppliers get maximal profits. Thus although we can generate the same, optimal, diffusion path of a technology under different supply regimes if expectations of buyers differ, the incentives to do R & D are very different under the two regimes and thus a welfare definition that also considered the generation of technology would not judge the two paths as equally desirable.[46] One might further note that the assumption that the new technology is domestically produced is crucial to the results.

(2) The linking of diffusion to R & D means that technology policies aimed at diffusion will affect R & D and policies aimed at R & D will affect diffusion, e.g. stimulation of a new industry through R & D support may make available cheaper or better products that stimulate faster diffusion. Similarly, stimulating use may improve incentives to R & D in the supply industry.

(3) The consideration that supply is also important leads one to argue that supply side reaction to demand orientated policies is an important

---

[45] These definitions are appropriate in a world in which markets clear. When markets do not always clear one might, for example, use employment as a welfare indicator.

[46] The implications of this are explored in Stoneman and Ireland, 'Innovation and diffusion'.

factor to consider in evaluating the potential effect of policies, e.g. if supply capacity is limited, stimulating demand may only lead to higher prices and not greater diffusion. Stoneman and David compare the reaction of a monopoly supply industry to the reaction of a perfectly competitive supply industry, when information and subsidy policies are used to speed diffusion.[47] It is argued that the monopolist may react to negate the intention of information policies and the policies may not work. A competitive supply industry will not so react and information policies will speed diffusion, but subsidy policies under competition may lead to use of a new technology by firms for which it is not profitable and this may reduce welfare.

(4) Finally, recent advances incorporating expectations into the diffusion process have shed new light on policy analysis. In particular, if subsidies are expected and if stimulation of technological improvement or extensions of use are expected, then these expectations may lower current adoption rates.

Analysis that incorporates these sorts of considerations is limited.[48]

Conclusions

In this survey I have attempted to move away from treating the problem of diffusion as an exercise in finding the right S-shaped curve to fit the data and I instead look at the underlying theoretical bases being discussed as foundations on which to approach the analysis, understanding and policy implications of diffusion processes. The literature I have surveyed often tends to be of a high degree of mathematical complexity but I have tried to minimize this in the survey. Moreover most of the literature I have surveyed tends to be rather neo-classical in sentiment. I do wonder if I have really done justice to the 'political-economy' literature. Having said this, however, it is clear that recently numerous advances in the theory of diffusion have been made, these suggesting that the phenomenon needs much more sophisticated treatment than has been common in the past.

---

[47] Stoneman and David, 'Adoption subsidies vs. information provision'.

[48] A variety of these issues are discussed in P.A. David and P. Stoneman, 'Will technology policy improve the diffusion path', mimeo (Stanford, Cal., 1984), and P.A. David, 'New technology diffusion, public policy and industrial competitiveness', paper presented at the Symposium on Economics and Technology (Stanford, Cal., 1985). They are also discussed more extensively in P. Stoneman, *The Economic Analysis of Technology Policy* (Oxford, 1987).

It is appropriate that I finish this survey with my own views as to where research ought to go from here. I suggest basically four potentially fruitful lines.

1   There is an obvious need for empirical (econometric) work looking at the new factors recently introduced into diffusion analysis, i.e. expectations, supply sides, Bayesian learning etc.
2   We need to have much greater understanding of the growth of capacity, number of suppliers and market structure development of industries producing new products. This will obviously interact with the R & D literature.
3   We need further work on the explicit modelling of *product* innovation and, associated with that, work on advertising and marketing. These latter issues have probably been treated more extensively by management scientists than economists.
4   Finally, there is an obvious need for much more work on the policy implications of the theory of diffusion. It is no longer just a matter of discussing instruments: our recent advances enable us to discuss optimality, divergence from optimality and policies to correct this. The cost is that this makes policy analysis more complex, but it also makes it much more interesting.

Notes on Contributors

Patrick O'Brien was formerly Professorial Fellow of St Antony's College, Oxford, and is currently Director of the Institute for Historical Research in the University of London, where he holds a chair in economic history.

Peter Mathias was formerly Chichele Professor of Economic History at Oxford and is currently Master of Downing College, Cambridge.

Maxine Berg is Senior Lecturer in Economics at the University of Warwick.

Gwynne Lewis is Professor of History at the University of Warwick.

John A. Davis is Professor of History and Chairman of the Centre for the Study of Social History in the University of Warwick.

Derek H. Aldcroft is Professor of Economic History and Chairman of the Department of Economic History at the University of Leicester.

Richard Whipp is a Senior Research Fellow in the Centre for Corporate Strategy and Change at the University of Warwick.

Volker R. Berghahn was formerly Professor of History at the University of Warwick and currently holds a chair in contemporary European history at Brown University, USA.

Paul L. Stoneman is Professor of Economics in the School of Industrial and Business Studies at the University of Warwick.

Index

Aberdeen, 25
Admiralty, 38
AEG, 104
agriculture, 15, 26, 46, 50, 51, 54–5, 86, 98
Aldcroft, D. H., 5, 134
Alès coal-basin, 66–82
Alfa motor company, 84
America, 4, 22, 61, 114, 121–41, 142–61; Latin America, 54; see also United States
American Motors', 125
Ansaldo engineering company, 101, 102, 103
Anzin coal mine, 76
Arkwright, Richard (1732–1792), 11
Ashley, E., 33
Ashton, T. S., 33, 34
Asia, 54
Austin (motor company), 122–4, 133, 138
Australia, 108
Automobile industries: 4, 84; British, 4, 110, 120–41; German, 148–9, 158

Bacon, Roger (1210–1292), 34, 36
Baldwin, R. E., 18
Banca Commerciale Italiana, 103
Barnet, C., 114
Barsanti, Eugenio (1821–1864), 100
beer, 20
Belgium, 74, 96
Benetton, 54
Berg, M., 4, 70, 79
Bergamo, 88–9
Berghahn, V. R., 4
Bernal, J. D., 33
Berthollet, Claude Comte de (1748–1822), 11, 15
Biella, 94–6
Birmingham, 30, 35, 38, 39, 40, 55, 59, 61, 63
Black, Joseph (1728–1799), 37
blacksmiths, 23, 39
blast furnaces, 20, 101
bleaching, 32
Bohemia, 97
Bologna, 87–8
Bosch, Robert (German engineering firm, Stuttgart), 148–50, 155

Index

Boulton, Matthew (1728–1809), 35, 39, 40
Boyle, Robert (1627–1691), 34
Breda, Ernesto (engineering company), 102–3
Brescia, 91
Bridgewater Canal, 24
British Leyland Motor Company (BLMC), 123, 127, 133–5
British Motor Corporation (BMC), 124, 133
British Motor Holdings (BMH), 123; see Leyland Motors, British Leyland Motor Company, British Motor Corporation
Brittany, 69
Bundersvereinigung der Deutschen Industrie (BDI), 160
Bundersvereinigung deutscher Arbeitgeberverbande (BdA), 160–1
Burchardt, L., 147

Calabria, 90, 91, 92
calico printing, 57–8
canals, 20
Cannadine, D., 46
capital formation, 46–9
car industry, *see* automobile industry
carding machines, 8
Carinthia, 92
Carlyle, Thomas (1795–1881), 43
Cartwright, Edmund (1743–1823), 8, 9
Castries, Maréchal Chales-Gabriel-Eugène La Croix de, (1727–1800), 70–82
Cento Bull, A., 93–4
Centralverband der Deutschen Industrie (CDVA), 151
Chapman, S., 46
Chaptal, Jean-Antoine Comtede Chanteloup, (1756–1832), 19
charcoal, 20, 91–2
Charnaron, J., 127
Chaussinand-Nogaret, G., 76
chemical industries, 15, 23, 32, 33, 35–6, 84, 100, 110, 112, 160
chemistry, 15, 34, 37
child workers, 60–4, 93–5
Chrysler (motor company), 123–4, 153
civil engineering, 36
Clapham, Sir John, 18
Clarkson, L., 65–6, 68, 74, 75
coal, 19, 22–4, 32, 57, 110, 112: coal-mining, 4, 22, 66–82, 110, 115
Coca-Cola Company, 153, 158
coke-smelting, 11, 21, 31
Cole, W. A., 44, 45, 47–8
Commonwealth, the British, 140
comparative productivity (British), 115–19
Cornwall, 40
Cort, Henry (1740–1800), 8
cotton industry, 45, 61, 91, 95, 100
Crafts, N. F. R., 44, 47–53, 56, 64
Credito Italiano (bank), 103
Crompton, Samuel (1753–1827), 11

Daimler (motor company), 121, 148, 150
Darby, Abraham (1677–1717), 8, 11
Darwinism, 'commercial' and 'technological', 27, 28; social, 146
Davis, N., 152

Dawes Plan (1924), 144, 153
Deane, P., 44, 47–8
democracy, 4
Derby, 87
design, 122, 132
Deutscher Metallarbeiter-Verband (DMV), 149, 151
Dunnett, P. J., 134
Dutch looms, 56–7

economic 'dualism', 51–6
education, 13; *see also* schools, science, technical schools, training
EEC, 114, 116–18, 128–9
electro-magnetism, 100
electro-technical industries, 104–5, 110
energy, 19–20, 23–4
engineering industries, 15, 22, 38–9, 98, 100–5, 110
experimental method, the, 15–16

Fairbain, Sir William (1789–1874), 11
Feinstein, C., 49
Fiat (Turin), 84, 105, 138
flax, 95–6
Floud, R., 47
Ford (motor company, US), 121–41
Ford, Henry I (1863–1947), 131, 153–4
Ford, Henry II (1917–   ), 131
Ford UK (motor company), 134
Fordism, 145–6, 153–5, 159
Foreman-Peck, J., 122
forges, 20
France, 3, 4, 13, 15, 18–19, 22, 23, 24, 34, 36, 66–82, 89, 96, 105, 116, 127, 142, 144, 149, 154
Friedrich, O. A., 160

fuels, mineral, 8, 21–2: combustible, 95

General Electric Company, 153
General Motors (motor company), 122, 125, 129, 131–3, 139–41, 153
Genoa, 97
Gensoul steam-heaters, 95
geology, 36
Germany, 4, 74, 96, 100, 104, 105, 109, 111, 112, 113, 116–17, 127, 135, 142–61
glass industries, 23
Goldin, C., 61
Gramsci, Antonio (1891–1937), 143–4
Grand'Combe mine, 68, 75–82
Guest, R. H., 135

Habsburg commercial policies, 92, 97
Hall, A. R., 33
handicrafts, 54–6, 56–63
Hargreaves, James (c1720–1778), 8
Harley, C. K., 44
Hartmann, H., 144–5, 155, 157–9
Hertner, P., 104
Hicks, Sir John, 33
Higgonet, P., 82
Hitler, Adolf (1889–1945), 156–7
Hobsbawm, E. J., 43
Hofman, P., 159
Holland, *see* Netherlands
Homberg, H., 147–8
Hong Kong, 19
House of Lords Select Committee on Overseas Trade (1985), 109
Huntsman, Benjamin (1704–1776), 11
hydro-electric power, 84

Index

IG Farben Chemicals Company, 153, 154
industrial relations, 122, 126, 135–8
innovation,
  contexts of, 3–5
  and economic growth, 107–10
  and economic theory, 5
  and military expenditure, 29–30
  single-cause explanations, 2, 118–19
  and technical change, 25–33
inventors, 12–13
iron and steel industries, 11, 20, 22, 24, 38, 67, 91, 100–1, 110, 115
Italy, 4, 54, 83–106, 138, 154, 161

Jacquard, Joseph-Marie (1752–1834), 11
Japan, 19, 127, 161
Jars, Gabriel (1732–1769), 15
Jevons, W. S. (1835–1882), 19
Johnson, C., 74

Kay, John (1733–1764), 11
Kennedy, John (1769–1855), 11
Kennedy, W. P., 111–12
Krupp engineering company, 101, 153
Kuznets, Simon, 6, 18

La Barberie, M., 71, 73, 81
labour, 2, 20, 22, 26, 30, 46, 56–63, 73–4, 93–6, 106, 110: *see also* children, women, productivity, skills, industrial relations, wages
Lancashire, 38, 40, 55, 59, 112
Lanchester (motor company), 121
Landes, D. S., 33, 71

Languedoc, 67, 68, 69, 75, 76
Leblanc, Nicholas (1742–1806), 11, 112
Levine, D., 61
Lewis, A., 18
Leyland Motors, 123
Lindert, P. H., 44, 49–50
Link, W., 144
locomotives, 24, 98, 102–3
Lombardy, 92, 93–5, 97
Lombe, John (1693–1722), 87
London, 20, 24, 38, 40
Lorenz, E., 115
Lotus motor company, 123
Lyons, 89

McCloskey, D. N., 44, 47, 111
machine-making, 27, 38, 39, 96, 97–8, 104
machine-tools, 27
MacIntosh, Charles (1766–1843), 11, 33
'Malthusian' perspectives, 52
management, 4, 115–19, 126, 131, 132–3, 136–41, 145–61
Manchester, 24
Marconi, Guglielmo (1874–1937), 100
Marshall, A., 18
Marshall Plan, the, 158, 161
Martin-Siemens furnaces, 101
Marx, Karl (1818–1883), 8, 73
Marzotto Woollen Company (Valdagno), 96
mathematics, 37–8
Maudslay, Henry (1771–1831), 39
mechanical-engineering, 38, 102–3, 138
Meier, G. M., 18
Merger, M., 101, 102–3
metallurgy, 15, 100
metalworking trades, 56–7

Midlands, 22
Milan, 86, 88, 99
mining, 24, 91–2: see also coal-mining
Mokyr, J., 46, 48–53, 56
Morris (motor company), 122–4, 133
motor industry, see automobile industry
Musson, A. E., 33

Nantes, Edict of, 67
Naples, 97, 101
Napoleon 1 (1769–1821), Emperor of the French, 71, 82, 90
National Socialism, 156–7
Nef, J., 45
Netherlands, 88
New Deal, the, 156
Newcomen, Thomas (1663–1729), 8
non-conformists, 13
Normandy, 69
Northampton, 33
Northumberland, 40
Nuffield motor company, 123: see also Morris

O'Brien, P. K., 5, 44
Olivetti company, 84
Opel motor company, 148, 150

Paris, 38
patents, 26, 40
patronage, 14
Paul, Lewis (d.1759), 8, 27
Périer brothers, 71, 73
Perrone, Ferdinando Maria, 102, 103
Piedmont, 74, 87–8, 92, 93–5, 97
Pirelli rubber company, 103

Poni, C., 87–9
production techniques, 137
productivity, 7, 38, 45–64, 110–19, 126, 129
proto-industrialization, 50, 65–82, 94–5

Racconigi, 88
railways, 20, 97, 100–1, 102–3
Ramella, F., 94
Red Indians, 82
Renault (motor company), 149
resources, and economic development, 18–21; and technical change, 21–5
rice cultivation, 86
Richard, Carrouge et Cie, 69, 75–6, 81
Riley (motor company), 121
Roberts, Richard (1789–1864), 8, 11
Robinson, A. E., 33
Roebuck, Dr John (1718–1794), 33, 37
Rootes (motor company), 122
Rosenberg, N., 109–10
Rossi, Alessandro (1819–1898), 96, 98
Rostow, W. W., 33
Rothwell, R., 119
Rover (motor company), 121, 123
Royal Society, The, 33, 38
Russia, 10, 15, 31, 34

S. Leucio silk works (Caserta), 89–90
Saito, O., 60
Savery, Thomas (1650–1715), 8
Schofield, R., 44, 60
schools, 13: see also technical schools, universities
Schott, K., 107

science, 2, 14–15, 19, 27, 33–9, 99, 100
scientific instruments, 38
'scientific management', 145–51
Scotland, 13, 23, 24, 37, 40
Scots, 13
Sheffield, 38, 55, 58, 59
ship-building, 30, 97, 110, 115
Siemens-Schuckert company, 104, 147, 155
silk industry, 62: in France, 3, 69, 72: in Italy, 87–91, 92–4
skill, 15–16, 22–3, 25, 32–3, 35–8, 39–40, 56–64, 99–100, 112
Sloan, Alfred, (President of GM), 131–2
small-scale production, 50–63
Smith, Adam (1723–1790), 29, 38
societies, scientific and philosophical, 15, 33–4
Sokoloff, K., 61
Solvay process (of alkali production), 112
Spalding, 33
spinning, mechanized, 3, 27, 31, 55, 94–5: jenny, 8, 56, 62
Standard motor company, 122, 123
steam engines, 8, 11, 27–8, 39–40; Savery and Newcomen, 8, 40
steam pumps, 73
steamships, 8, 20
steel industry, 24
Stevenson, George (1723–1790), 40
Stoneman, P., 5, 107–8
Sweden, 31, 56, 138
Switzerland, 88

Taylor, F. W. (1865–1915), 145–6
Taylorism, 145–51

technical schools, 99, 100–3
techniques, 11, 25
technologies,
American, 4, 116, 142–61
definitions, 2–5, 6–10
demand-based theories of, 10–12, 138, 164–77
'high' and 'low' technologies, 3–4, 27–9, 51–63, 84, 110–15
public policies and technology diffusion, 181–3
research & development and, 180–1
and science, 12–16, 33–41
'supply'-based theories, 12–16, 177–80
transfers of, 143–61
Terni steel works, 101
Texans, 25
textile industries, 3, 15, 20, 23, 27, 31, 45–6, 56–9, 61–3, 87–91, 92–6, 110, 112
thermo-dynamics, theory of, 14
Tosi, Franco (engineering company), 102
training (technical), 13, 100–3, 159: *see also* TWI
'Training Within Industry' (TWI), 159
transport costs, 19–20, 23, 25
Treviso, 85
Trevithick, Richard (1771–1833), 8, 9, 11
Triumph motor company, 121
Tron, Andrea, 85
Tubeuf, Pierre-François (1738–1795), 70–82
Tucker, Josiah (1712–1799), 58
Turin, 88, 99
Tyszynski, H., 113

United States, 22, 25, 61, 63, 111,

United States – *cont.*
    114–16, 120–41, 127, 129–41, 142–61
universities, 15, 37, 99–100, 104–5
Usher, D., 107

Vandermonde, Alexandre (1735–1796), 15
Vaubel, L., 160
Vaucanson, Jacques de, (1709–1782), 11
Vauxhall motor company, *see* General Motors
Verband Deutsche Ingenieure (VDI), 146–8
Vinci, Leonardo da (1452–1519), 10, 28
Virginia, 80

wages, 31, 60, 73–4, 96, 106, 107
Wales, 24
Warrington, 59
wars: Crimean War, 36; First World War, Great War, 4, 84, 105, 106, 122, 151–2, 156; Second World War, 4, 53, 114, 118, 122, 133, 139, 157
watch-making, 38

water-power, 22, 30
Watt, James (1736–1819), 8, 9, 11, 27, 31, 37, 40
weaving, hand-loom, 3, 31, 55, 94–6; mechanized, 27, 95–6; *see also* Dutch looms
Weimar Republic, 156, 161
Wells, L. T., 128–9
Whipp, R., 4
Wilkinson, F., 115
Wilkinson, John (1728–1808), 30, 39
Williamson, J. G., 44, 46, 48–9, 56
women workers, 4, 56–64, 93–5; productivity, 4, 58, 60–4; *see also* skills
woollen industries, 61, 94–5, 96
work practices, 50, 52–3, 56–64, 73–4; in the car industry, 122, 135–41
Wrigley, E. A., 44, 52, 56, 60–1

Young, Arthur (1741–1820), 86

Zangen, W., 155
Zegveld, 119
Zeiss Opticals Company, 153
Zeitlin, J., 136

Lecture Notes in Computer Science 663

Edited by G. Goos and J. Hartmanis

Advisory Board: W. Brauer D. Gries J. Stoer